Claire O'Rourke is a campaigner, communicator, behaviour change expert, partner and parent. Totally climate freaked-out, Claire helps people and organisations take action on climate change, currently as Energy Transformation Program Co-Director at The Sunrise Project. Previously Claire was National Director of Solar Citizens, a community-led renewable energy advocacy organisation. A former journalist, Claire has extensive experience campaigning for social impact, including driving communications for the Every Australian Counts campaign for the National Disability Insurance Scheme and as a senior leader at Amnesty International Australia.

Half of the royalties from *Together We Can*
will go to Groundswell Giving:
www.groundswellgiving.org/

PRAISE FOR
TOGETHER WE CAN

'If you ever thought you were alone in the battle for climate justice, this book will make you realise you are part of an extraordinary movement of people.'

Craig Reucassel, ABC TV's *The War on Waste*

'Claire O'Rourke shows us thoughtfully and thoroughly the impact climate change is having on our mental health as individuals and communities, and the importance of hope in facing these current and future challenges.'

Daisy Turnbull, teacher, Lifeline counsellor and author

'This book reminds us that with nothing more than hope and action, ordinary people do extraordinary things.'

Kelly O'Shanassy, CEO, Australian Conservation Foundation

'This book is the hope-filled, forward-thinking, realistic and *practical* antidote required for the days when climate anxiety threatens to derail us. A reminder that no matter how overwhelmed we may be feeling, there is always something we can do, those efforts matter, and we are never, ever alone. It's a fortifying, inspiring book to return to again and again as we navigate the messiness of creating a better way forward.'

Brooke McAlary, author of *Slow*

'Exactly what we badly need right now and I learned a lot. This incredible book shows how we can find deepened connection with others while getting out of our fossil trap.'

Ketan Joshi, author of *Windfall*

'The answers to climate catastrophe are here now, and *Together We Can* is a testament to the people building solutions everywhere.'

Danny Kennedy, CEO of New Energy Nexus and co-host, *Climate of Change* podcast

'The journey from "climate freak out" to "cultivating hope" is re-imagining our future. This book can help you change your mindset . . . and indeed your world.'

Dermot O'Gorman, CEO WWF-Australia

'The ultimate companion for the climate age. Full of inspiring stories about climate action for all of us who feel overwhelmed by global warming.'

Rebecca Huntley, author of *How to Talk About Climate Change in a Way that Makes a Difference*

'A campaigner lies asleep inside you; *Together We Can* wakes it up.'

Ashjayeen Sharif, student climate activist

'O'Rourke's book is pragmatic, hopeful and richly democratic, showing there are multiple pathways for every willing Australian to get involved in driving our clean energy transformation at emergency speed and scale.'

David Ritter, CEO of Greenpeace Australia Pacific

'Only a vast array of actions, big and small, can defuse the climate crisis. This book shows you how.'

Professor Will Steffen, Australian National University

*Everyday Australians doing amazing
things to give our planet a future*

TOGETHER
WE CAN

CLAIRE O'ROURKE

ALLEN&UNWIN
SYDNEY·MELBOURNE·AUCKLAND·LONDON

Allen & Unwin
83 Alexander Street
Crows Nest NSW 2065
Australia
Phone: (61 2) 8425 0100
Email: info@allenandunwin.com
Web: www.allenandunwin.com

A catalogue record for this book is available from the National Library of Australia

ISBN 978 1 76106 681 8

Internal design by Samantha Collins, Bookhouse
Set in 11/17 pt Stempel Schneidler by Bookhouse, Sydney
Printed and bound in Australia by Griffin Press, part of Ovato

10 9 8 7 6 5 4 3 2 1

The paper in this book is FSC® certified. FSC® promotes environmentally responsible, socially beneficial and economically viable management of the world's forests.

For Martin, Maeve and Illona, my universe.

CONTENTS

This book was written on Dharawal Country, land that was never ceded and a place I am privileged to call my home. I respect and honour the traditional owners of this land and Elders past, present and emerging today, those who we must learn from to ensure our planet thrives. Always was, always will be.

INTRODUCTION

Tell me, how are you *really* feeling about the state of the planet? Are moments of worry, sadness or exhaustion creeping up on you when you least expect it? Maybe your brain is burrowing down wombat holes and you're consumed by thoughts that anything you can do to help address climate change will only scratch the surface. Perhaps you can't decide what to do or where to start and feel a little overwhelmed. My friend, these are the common symptoms of climate freak-out, a term that I find best describes the mixture of all these swirling emotions: the gnawing grief about the loss of our world's natural systems and species; the fear of the future coming at us far too quickly; the fury at the slow pace and scale of action from our representatives; the hopelessness of a problem too big to solve.

Climate change is, after all, enormous and pervasive, which is why it's understandable that you or those close to you could be caught in these emotional responses from time to time. Let me reassure you that millions, yes *millions*, of people are feeling the same way: eco-anxiety is fast becoming a business-as-usual area of psychological study because it's just so prevalent. There's plenty of bad news around, so you may not be shocked to hear that climate freak-out is on the rise, but you might be surprised to learn of the sheer scale of the positive actions that millions of Australians like you and me are taking to transform our country and heal our planet, right under our noses. They're inventing new technologies, building circular economies, restoring forests, creating new community connections and reinventing our food systems. It's time to take a moment to look up from the online vortex of frighteningly bad news and see what we can do to heal ourselves and our world.

I'm in an unusual line of work: supporting people to advocate for big climate-policy changes by governments and to push for the huge shifts needed from businesses to decarbonise. I'm privileged to be up close every day with passionate, caring people who are on the front lines of a vast array of social, economic and environmental actions. Through this work I've seen many cases of freak-out and burnout, and I was there for a while too before I found a way to get moving again.

It was Australia's bushfire disaster of summer 2019–20 when I hit a wall. There are a few flashes of memory from that time: day after day of compulsively doom-scrolling the news/emergency services/weather forecast/air-quality apps; fixated thinking about how all the places I loved and had brought my children up in would be gone; my pre-teen daughter in tears as we packed the trailer with evacuation kit in 40-degree heat; eerie smoke-hazed skies full of thousands and thousands of endangered flying foxes circling at dusk, something we'd never seen before. It dawned on me that my family would experience more of this: the extreme heat, longer fire seasons, more frequent drought, shortages of food and water perhaps. The fear landed with the realisation that my wooden home in the trees would likely burn, if not this summer, then one summer to come soon. Societal collapse in our stable democracy seemed all too possible, and it was actually a shock. My usual positivity faded fast and my motivation dropped overnight. I kept up a veneer of energy at my job—yes, my *climate advocacy* job—with my partner and our children, with my friends too, but it was pretty rough. I had to get out of this rut.

I began by looking into research around how to cultivate a little more hopefulness in my worldview. Hope is, according to experts, *necessary* for human resilience. It turns out that hope isn't a genetic optimism index; that is, you're not born with your glass half full or half empty. People are taught how to hope, if they are lucky, by their parents when they are young. Who knew? If hope can be learned, logic told me that we must be able to build our hope muscles through training of some sort. Like any transformation program at the gym, I reasoned it was going to have to start with a commitment: a decision to change. So as the late summer rain finally started falling on the tin roof and the smoke faded from the sky, my climate hope project began. I decided to fill my cup of optimism by taking a closer look at the actions that individuals, families, communities, workplaces and industries are taking across our nation, and share what I've learned along the way.

Let's be clear. This book is not about flaky, glass-half-full, ignorance-is-bliss denialism. It's not all climate-fixing rainbows and unicorns either: that would be just as annoying. Not every story you'll read about is a victory: many are of progress and experimentation rather than perfection; others are complicated, just like every one of us. The choice I made and I hope you make too, is to rise above what I see as the darkest consequence of climate panic: a drop in engagement and motivation that will lead us nowhere. The consequences of climate *in*action are just about as frightening as the scale of the problem we are facing. There is an enormous risk that we will turn inwards, disconnect from our communities and society and let everyone else worry about themselves. That's the kind of isolation and distrust that could tear us apart in the very moment we really need to come together to work on climate change in every way possible.

You may never have experienced a climate freak-out moment, but even so, *Together We Can* aims to show you a small sample of the energising news that is overlooked too often by climate doom-feeds. You'll read about how you can choose hope over fear, face climate change with confidence and jump in to face this challenge, wherever you are.

HOW TO USE THIS BOOK

This book will teach you how to build hope into your everyday thinking, and how it will empower your work, your family and friends, and your community. *Part 1: Totally freaked out* will show you how you can wrestle with the anxiety, fear and grief that climate change is triggering for people all over the country, and how social tipping points are generating positive impacts. *Part 2: Project, not panacea* demonstrates how you can reframe the climate challenge while not shying away from it, and explains how big business and even pesky politicians are stepping up. *Part 3: We've got this* presents useful frameworks you can use to identify the unique personal characteristics that you can contribute to your communities, and that you can use to learn how to master endurance from those with decades of climate action experience.

If you've picked up *Together We Can* it's likely you're already doing a lot to help the planet: thank you for everything you do. But this book will also get you thinking about the systems we need to shift and your role in influencing them. That's a big thing, and it may feel overwhelming at times, so the end of each chapter provides key takeaways for you to remember or refer to: think of these as the most important things you'd mention when chatting to a friend about the state of the world over a cuppa. I've included examples, ideas and entry points for taking action that build hope, bring connection and help to heal our world. What's more, on my website you will find the biggest list I could build of inspiring groups, organisations and movements that you can join or that might inspire you to start something with people who care as much as you do: see Climate Action Starts Here, at the back of this book.

This project was sparked by my climate freak-out, and it's led me to celebrate the vast (and it is vast!) community of people who teach me every day how to be hopeful. I've learned how important it is to cultivate hope with intention *and* action. My biggest hope is that you will read *Together We Can*, get inspired, make a clear choice to be hopeful, and build your plan to create a better, stronger and more resilient world, together.

Part One

TOTALLY FREAKED OUT

CHAPTER 1

THE MOMENT OF TRANSFORMATION

'Do not lose heart. We were made for these times.'

—DR CLARISSA PINKOLA ESTÉS

Wander a short way down the Bermagui Road and you'll find yourself on the manicured lawn outside the Cobargo School of Arts, the beating heart of this small village on Yuin Country in Northern Bega Valley Shire, New South Wales. More than a century of community memory is preserved in the hall, its tidy weatherboard and stone walls bearing witness to scores of weddings, funerals, concerts and fundraisers in years gone by. With high ceilings, sash windows, eye-numbing fluorescent lighting and a tiny stage well-suited to hosting performances of SOAPI, the School of Arts Players Inc. amateur theatre group, it is a simple, spartan place, all practicality and no fuss, like the folks who live here. Here, on a hazy summer evening, a traumatised community took a deep, nourishing breath.

Back on that February evening, Cobargo's tiny township of 800 people was beginning to emerge from a summer of repeated evacuations, searing heat and suffocating smoke, the symptoms that were the hard, familiar hand of that year's bushfire season. New Year's Eve was the longest night: with only twenty minutes' warning, the town was overrun by the Badja Forest Road Fire, the most destructive blaze in the region's living memory. Cobargo was just one of the many coastal towns set upon by the bushfires that raged during Australia's Black Summer. Four lives were lost, 108 homes destroyed and ten Cobargo businesses wiped out on the evening that brought in the new decade: the School of Arts would have burned were it not for a canny neighbour who leapt to its defence. Years on, ridgelines surrounding the town carry the ragged, bare bones of the land, a reminder of those long nights battling a landscape under so much pressure it broke down.

Local celebrant and community volunteer Debra Summer woke in the early hours of that morning with an asthma attack and, reflecting on her experience of the Canberra bushfires almost two decades earlier, decided to keep vigil. Something, she says, 'just didn't feel right' that night. 'I saw the flames coming across very quickly, and so I woke everyone up and to cut a long story short—it was all chaos—we decided that we needed to evacuate with the kids,' she said. 'I think we were possibly the last ones to make it out of Cobargo before the road was blocked. There were flames everywhere you looked, we were driving past flames everywhere.' Debra's husband David Newell made a last-second decision to stay behind to defend their home, his brother-in-law jumping out of the car to help. 'It was pretty terrifying leaving them behind, not knowing if they were safe.'

The experience was hardly unique in the 2019–2020 bushfire season, with more than 24 million hectares of the country burned, an estimated 3 billion animals killed, injured or displaced and 3000 homes destroyed. Thirty-three people lost their lives, in events that stunned us with their ferocity and relentless destruction as the weeks pushed on. For the first time climate change was widely acknowledged as the culprit of the calamity, after decades of warnings from scientists that Australia would experience more frequent and intense storms, droughts and bushfires. Cobargo itself was etched into Australia's national psyche when footage of burnt-out homes and shell-shocked locals was beamed around the nation, a visiting prime minister taking hefty serves of outrage from a community on its knees. Only a few weeks after the peak of the crisis, on that Monday evening in February 2020, when the destruction was fresh and many consequences of the trauma were still unknown, Cobargo began a lesson in connection and healing that deserves a closer look.

Cobargo's community has competence and 'can-do' woven into its DNA, with a long history of volunteering in the fabric of the people here. With so many friends and neighbours raw and hurting, Debra Summer and husband David, with Cobargo Folk Festival Director, Zena Armstrong, decided to have a go at something a little different to the business-as-usual community consultation playbook, which all too commonly sees outsiders sweeping in to run the show in the weeks and months following a disaster. 'We were thinking we need to start pulling together to start visioning, co-designing a way forward as a community, so we just decided to start these Cobargo Community Catch-ups,' Summer said. Without knowing if anyone would show up, word was sent around and thick plastic school chairs were arranged in small circles around the hall. 'The first meeting ended up with well over 100 people,' Summer said. 'It created an opportunity for people to share their stories in a safe space.'

Scott Herring walked through the hall's front door that evening because he wanted to be useful. 'It was like a real touchpoint for people, especially for people who felt that they had something to offer or wanted to contribute,' he said. It was remarkable that Herring turned up at all: his family was left homeless when their home and shed on 40 hectares were destroyed, and he was in the middle of a heavy firefighting season in his job with the National Parks and Wildlife Service. Reflecting on that night he remembered the emotional intensity that flared in a small group conversation: it was serious business. 'I realised that this is not to be taken lightly, what we're doing here; we're not here for good feelings and self-satisfaction. We're here because people are really being smashed, and people need help.' Before long, the hall started humming with ideas among the laughter and tears, and the village's recovery engine got the jump-start the community craved. The extent of the change the gathering made to people's emotional state is shown in the box 'Check-out'.

HARVESTING HOPE

Debra and David are practitioners in the Art of Hosting, a method of guiding group conversations that aims to encourage self-organisation of groups, so they can quickly respond to challenges, opportunities and change. Based on a philosophy that wisdom is already present in a group, the 'art' is in surfacing, collecting and reflecting back ideas and strengths so that the community can harness them. It's worth reflecting on the exquisite simplicity of the hope that can flourish at times when humans become deeply and tangibly connected, moments that are all too rare in our frenetic daily lives.

CHECK-OUT

Cobargo's residents described how they felt when leaving their first Community Catch-up night in February 2020:

- Encouraged
- Lighter
- Open
- Curious
- Connected
- Love
- Willingness
- Grateful
- Better
- Hopeful
- Inspired
- Included
- Belonging

In Cobargo, the gatherings planted the seeds of what has become an organic and evolving network of projects to help the community flourish again. At that first gathering, people were asked what they noticed in the community since the fires. Shock, anger, despair, fractures, isolation and dislocation bubbled up, but so did independence, self-empowerment, generosity, dreaming and hope. This key question prompted more considered responses: 'In the aftermath of the fires, how can we best work together for our wellbeing and the future of Cobargo?' It was the beginning of many more catch-ups, where the community came to generate ideas, build networks and get recovery happening on their own terms. (See the box 'Our principles for working together'.)

OUR PRINCIPLES FOR WORKING TOGETHER

The Cobargo Community Catch-up's Principles of Working Together were decided at the first meeting, in February 2020:

* Listen with attention.
* Speak with attention.
* We invite collective wisdom.
* Ensure every voice is heard.
* Offer what you can; ask for what you need.

Zena Armstrong, also the chair of the Cobargo Community Bushfire Recovery Fund established after the fires, said the catch-ups were important in establishing common interests, and prioritising what was most important to nurture as the community began to heal. 'There's a tremendous amount of insight, new knowledge and new thinking coming forward,' Armstrong said. 'We've been able to draw this together as a community and it's very exciting.' Themes coalesced around energy, food security and more shared resources: underlying all of this was the importance of maintaining community connectedness while preparing for a more uncertain future. 'We have taken advantage of this terrible thing that has happened to us, to practical effect,' Armstrong said. 'We've taken a whole bunch of aspirations that were already present in the community that we may have got round to pursuing over time, but this galvanised us. It encouraged us to have confidence in our own capacity to lead our recovery and to develop some very significant community-led projects.'

Cobargo is rising together with competence, care and connection from this climate change–fuelled catastrophe to build a more sustainable home base. There is no doubt this community's trauma runs deep, with more lives lost in the months after the tragedy: the impacts on physical and mental wellbeing persist in this place, like so many around Australia that were trampled by that long, long summer. But in the months since that first meeting and in the context of the Covid-19 pandemic, a jaw-dropping number of recovery efforts have transpired. Around $800,000 raised through the bushfire recovery fund in a collective effort spearheaded by the Yuin Folk Club has enabled many of the ideas floated at the gatherings, helping with seed funds that eventually secured close to $18 million to rebuild the town centre and to build a solid case for a solar-powered micro-grid for the village.

Scott Herring brought the idea of a tool library to the catch-ups, and it is now a hub of community support—there is some very impressive-looking heavy-duty kit available to borrow. The community garden has a permanent home, the Cobargo RFS has new critical fire-fighting infrastructure, the village has well-developed plans for a new community centre/disaster refuge, local schools and pre-schools have rebuilt bushfire-damaged play areas, and a plethora of community art and wellness projects are rolling out, including Ginger the Frog, a creative wellbeing program for kids (and adults). The program is produced by Sarah Campbell Lambert, founder of the Cobargo Wellness Group that also began its life in early 2020. 'What supported moving us forward was connecting those groups, and new groups that formed, so people knew what was going on in the town,' Summer said. Cobargo's idea of contributing with a little time, space and care feels tantalisingly straightforward in this fast-paced, technology-infused world full of complex, seemingly intractable problems. People can and do have a voice.

Author Greg Jackson wrote of the challenge of organising ourselves around the climate crisis before its worst consequences hit, calling for us to 'reknit' the fabric of shared purpose, with helping, sharing, protecting or teaching, all enlarging human experiences. 'It forges stronger relationships, infuses purpose in one's actions, and affirms the deepest seat of worth a person can possess: having something to contribute that matters to other people.' The challenge that climate change presents will test us, individually and collectively. It will push our society and our humanity to the brink. What could we achieve if we stepped forward now, not later, to create human connection and goodwill that transforms lives, relationships and communities? Could we rebuild our society so we restore the damage we've caused *and* create a world where we can all flourish? What would this mean for our planet? What vision might we realise?

TOTALLY FREAKED OUT

My life feels pretty average-normal Aussie in so many respects. I'm a partner to my sweetheart from university days and a mid-forties mum of two teens and Luna, a jet-black rescue labrador who barks far too much at passing dogs, the postie . . . everything and everyone really. We live in a house wedged between rainforest and the sea, which means carrying a mortgage that's a little too high to feel terribly secure. My days flow with a torrent-like quality that makes high stress an everyday pastime. Rest is an indulgence and life logistics leave my calendar a game of Tetris, trying to cram everything into a seamless pattern: jamming in work, dinner, exercise, calling friends, remembering (forgetting) birthdays, planning holidays, saving money, paying bills, driving the kids around and whatever else I've left to linger on a long to-do list. All this is the usual flavour of the weeks that accelerate into months, then years. I suppose this is why I'm always shocked when a moment of climate freak-out interrupts my flow, making me draw a sharp breath. It's those split-second pauses that are brought on by the run-of-the-mill daily news, when you see another species' extinction ranking lower than the latest political bunfight over submarines or sports rorts, or when my daughter says with utter conviction that she will never have children. The moment coasts along my shoulders, sometimes landing with a tightness in my throat or casting a shadow over my eyes. When I read that the mayor of the Spanish city of Seville, Juan Espadas, announced the city would be the first to begin giving personal names to individual heatwaves, officially recording them like a cyclone or hurricane in the history books, goosebumps ran down my arms in seconds, my hair actually standing on end. So many times I've thought, 'So this is how it starts.'

These thoughts and feelings—mixing anger, sadness, anxiousness and fear—are such strange interruptions to delicious moments of average-normal life, I didn't feel like I could really talk to anyone about them, let alone face them. I'd had a scare over the fire season, but I was safe and secure, my life of privilege was intact, so I thought I should push these feelings aside. Besides, most folks I admire are gritty climate-justice warriors, so I felt I was probably one of only a few people feeling this way. But I was as surprised as anyone to learn just how common these feelings about the state of the world are. They're often prompted by increased climate impacts like fires, heatwaves, intense storms, flooding and drought that we're experiencing or anticipating: in January 2020, at the peak of the Black Summer fires, research from The Australia Institute found that people who considered themselves 'very concerned' about climate change jumped from

37 per cent to 47 per cent, compared with only six months earlier (see Figure 1.1). It's one of the fastest and most significant shifts in public opinion I have observed in two decades of advocacy and campaigning work.

Was there a collective, nationwide trauma here, a surge of interest and concern that would ensure we put the enormous shift required to rapidly decarbonise within our reach? A group of committed advocates (myself included) and academics decided to find out, and within a few months, as the whisper of a mysterious

FIGURE 1.1 HOW CONCERNED ARE YOU ABOUT CLIMATE CHANGE? (JULY 2019 TO JANUARY 2020)

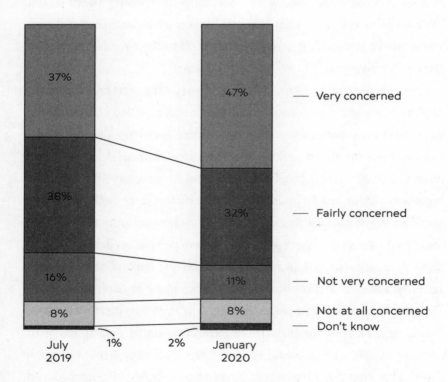

Source: The Australia Institute

coronavirus became a Covid-19 roar, a new research project known as Climate Compass landed in the world. As it turns out, if you're feeling freaked out about climate change like I am, you're in good company: there are many, many worried people across our nation.

Climate Compass, a survey of 2500 Australians conducted in July 2020, provided an analysis of attitudes to climate change, which found 24 per cent of Australians settle nervously into the 'Alarmed' group—that's around 4.9 million people aged 16 to 75 years old. People in the Alarmed category are *really* worried about climate change, so much so that 76 per cent of people in this group name climate change as their top policy area, followed by, you guessed it, the environment, which rates at 64 per cent. The vast majority of this group (80 per cent) believe humans are fully responsible for climate change and understand the harm to current and future generations that will come. And when you look at young people aged 16–24 years old, the numbers don't lie: the proportion who are Alarmed jumps another six points to 32 per cent.

A national survey on climate change and mental health conducted by researchers at Deakin and Monash universities between August and November 2020 found, from 5483 people surveyed in Australia, that eco-anxiety is running at significant levels in younger age groups: at 23.5 per cent among 18–24 year olds, 19.6 per cent among 25–34 year olds and 13 per cent among those aged 35–44. 'It was across the board, from rural and regional areas to cities,' said research lead Dr Rebecca Patrick, from the Health, Nature and Sustainability Research Group at Deakin University. Post-traumatic stress disorder was running at 25.6 per cent among people who had direct experience of a climate-related event like bushfires, floods and drought. Nearly 16 per cent of participants who had been spared from experiencing such events met the screening criteria for pre-trauma stress, a finding Patrick said highlights the significant and enduring impact of anticipated climate events on the mental health of the general population. A separate analysis based on the same survey also found two-thirds of respondents, or 66 per cent, reported that climate change was a personal concern, well ahead of Covid-19, which sat at 25 per cent, despite the survey taking place when Australians were experiencing escalating stages of lockdowns in several locations. So many of us are waking up to the reality of living in a world that is radically changing within only one generation, and it's a pretty scary prospect to consider what's coming.

THIS CHANGED EVERYTHING

Welcome to Kingaroy, only 212 kilometres north-west of Brisbane, but just about as far away as possible from the concerns of the chardonnay- and latte-sipping crowds of the inner cities. Known as 'red ant' in the local Wakka Wakka language, the town was famous for decades as the home of long-time premier and prime ministerial aspirant the late Sir Joh Bjelke-Petersen. Navy beans, maize and sorghum grow well in Kingaroy, but the town is even more famous for being the peanut capital of Australia: after 50 years of locals talking about it, a 450-kilogram steel peanut was finally raised in November 2021 at the local Lions Park to recognise the title. A pig slaughterhouse, Swickers, is a major employer, and the Meandu coalmine and Tarong power station are only a few clicks down the Kingaroy–Cooyar Road. Call me a captive of my formative adult years crawling inner-city Sydney, but it seemed pretty unlikely that this would be the place you'd find climate freaked-out folks. That was what I thought until I met Suzanne Mungall.

Mungall hit her mid-forties with the same enthusiasm she'd always had for life, thanks to a combo of four kids, a nice bloke, a great job and good health. The self-confessed life of the party is connected to everything local, from the soccer club to playing flute (badly, she says) and singing alto in the choir at the local Catholic church. Mungall is connected through the schools and hospitals across the town through her work as a speech pathologist and has even been involved in organising Baconfest, a crackling annual Kingaroy celebration where you can get a serve of bacon fries and bacon ice-cream before (or maybe after) you've had a go at the 'Rashers Rush' cycling and running event. A daughter of cattle farmers, Mungall wistfully described her idyllic rural childhood spent helping fix fences and manage vegetation on the family's Mount Morgan property in central Queensland. There wasn't a thought about the precarious state of life, the planet and all the rest. 'I thought I was doing the right thing, you know, this is good work, we're in nature, we're supporting the cattle. We loved our cattle, we knew every single one of them by their faces,' she said. 'How could that be wrong?'

So many things changed for Mungall over twelve short months. Studying a master's in public health planted the seeds of climate awakening; her son was waging a war on waste guided by his primary school teacher; and then, she mentioned, almost in passing, that more than 90 per cent of the family farm burned in a fire that swept across her property (thankfully her home was spared). But this swirling, unsettling year had a more unforgettable moment for Mungall when she was watching a documentary suggested by her son's schoolteacher, starring Greta Thunberg, the

Swedish teenage climate action juggernaut. Thunberg's narrowed eyes and fierce words resonated so deeply that a switch flipped in Mungall's head. Here was this European youngster with long plaits, a similar age to a couple of her own brood, pointing straight at her, calling out what she had not done, not seen or understood. 'When I first truly looked hard at the environmental stuff I was so shocked to the core, I remember feeling it was like perceiving God is not real,' she said. 'I was like, do you mean to say I have been raising cattle? I was the person out there killing all the trees!' Mungall was rattled, really rattled. The months passed and Mungall became more irritable, crying often and dreaming of the environmental devastation that was to come. She didn't panic, but the gloom certainly settled in, her eyes misting over as she remembered how she kept asking herself why she'd had so many children.

'It's like you're kind of constantly reminded, everywhere you look and in every move you make just living in the world that we live in and the way it's set up, you're kind of a bit trapped by it,' Mungall mused. 'You're at Woolies, and the only vegetables you can buy are packed in heaps of plastic and you know they've been refrigerated with some enormous energy use for months and you feel like, well, what are the real options?' Considering the scale of everyday living, Mungall came to a realisation that's hard for anyone to bear: that we're stuffed, totally stuffed. 'I remember I googled "eco-anxiety" and I was on the money with that word and the information came forth, but in a way it helped less because I couldn't just pretend I was the stupid one. I would have liked to have read the opposite, that this is irrational. In a way it was worse because it was real.' One day, she decided to start talking to her friends about it, perhaps to relieve some of the inner turmoil, perhaps as a cry for support. 'We sat in the coffee shop, and I was just bawling my eyes out.'

Suzanne Mungall is not alone. Leafing through the Climate Compass results, I keep drifting back to the responses to a simple open question: 'How do you feel about climate change?' The emotions people self-reported makes for arresting reading that shows the nature of the emotional challenge so many of us are facing. Hopeful optimism isn't exactly winning the race, with a colossal 82 per cent of respondents nominating a plethora of difficult, charged emotions (see Figure 1.2).

Scrolling through the answers from participants reflects a disturbing reality that so many people are experiencing. Chances

FIGURE 1.2 TOP 10 NEGATIVE FEELINGS ABOUT CLIMATE CHANGE

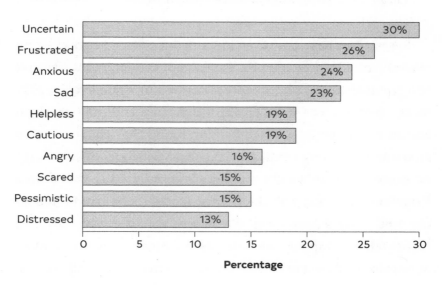

Source: Climate Compass Audience Segmentation Research, The Sunrise Project/ FiftyFive5, July 2020

are, if you're reading this book, some of these words could feel all too familiar:

'I swap between anger and despair.'

'It makes me feel sad, frustrated and angry that we have done this to the very place we live in. And seemingly still have no concern for the other inhabitants of this earth. What will we leave future generations?'

'Climate change makes me very scared, because I don't know how exactly our climate will change and how this will affect us. We only see its effects as it plays out.'

'Climate change is scary for me. I would like to buy mainly organic products, but they are expensive. I would like to use solar energy, but my property is rented. I am very worried about animals and nature.'

'Dread. I fear for my children (when I have them).'

Researchers have now even identified specific climate change–related emotions (see the box 'Learn your eco-emotions') as part of a groundbreaking global report. This report, completed in 2021 by the Institute of Global Health Innovation and the Grantham Institute at Imperial College London found that climate-related anxieties and strong emotional responses were identified in large proportions of university students and children in the United Kingdom, Australia and the United States (between 44 and 82 per cent). Take a breath before you read this: the report found a robust relationship between increased temperatures and number of suicides, and clear evidence of severe distress following extreme weather events. Climate change was found to exacerbate mental distress, particularly among young people, even for individuals who are not directly affected. The report also found people who

LEARN YOUR ECO-EMOTIONS

Emotions	Definition
Eco-anxiety/ eco-distress	Defined by the American Psychological Association in 2017 as 'the chronic fear of environmental doom'. May include anxiety, worry, stress, hopelessness, sleep disturbance, irritability, despair and symptoms of anxiety like upset stomach, awareness of heartbeat, shortness of breath and sweaty palms. Observed across all age groups and may especially affect young people.
Climate grief/ ecological grief/ eco-grief	This is grief felt in relation to actual or anticipated experiences including loss of species, ecosystems and meaningful landscapes. Found to affect those with strong ties to a particular place and people who witness environmental destruction, such as Indigenous and farming communities, and climate scientists.
Solastalgia	The term (coined by Glenn Albrecht) for the distress experienced when someone's home environment changes in negative ways, through climate change or through other causes, that attack someone's sense of home, creating a deep sense of homesickness. Reported in youth in Indonesia, Inuit communities in Northern Canada, farmers in the Australian wheat belt and communities around the Great Barrier Reef.

Adapted from Emma Lawrance, Rhiannon Thompson, Gianluca Fontana & Neil Jennings, *The Impact of Climate Change on Mental Health and Emotional Wellbeing: Current evidence and implications for policy and practice*, briefing paper, Grantham Institute—Climate Change and the Environment, Imperial College London, May 2021 meet criteria for mental illness are more vulnerable to the effects of climate change on their mental and physical health. If you're not triggered by any of that, how about this: the impacts uncovered by these royally smart scholars are 'likely to be vastly underestimated as despite the serious effects, this has been a neglected area of research'. What's more, until the 2021 meeting, mental health had *never* been a subject of discussion in any official side events at the Conference of Parties (COP) of the United Nations Framework Convention on Climate Change to date. Yikes.

The lead author of the report, Dr Emma Lawrance, chatted with me on Zoom from her family's quiet Adelaide Hills home, where she was having a pandemic-prompted break from her pad in bustling London. 'We hear from mental health professionals that there are cases of people coming to see them who are experiencing very extreme distress around climate to the point of some people even having some suicidal thoughts,' Lawrance said. 'This relationship between broader mental health concerns and these experiences of distress related to climate is still something we need to fully understand, but it's clearly acting as a stressor for people in their lives.' Young people are a particularly vulnerable group. A 2021 study of 10,000 children and young people aged 16–25 across ten countries (Australia, Brazil, Finland, France, India, Nigeria, the Philippines, Portugal, the United Kingdom and the United States) collected information on thoughts and feelings about climate change and government responses to the problem. The results demonstrate how widespread concern is among this age group. The research found 59 per cent of those surveyed were very or extremely worried: 53 per cent of respondents from Australia were either extremely or very worried about climate change, and another 29 per cent were moderately worried. Each of the following feelings—afraid, sad, anxious, angry, powerless, helpless or guilty—were reported by more than 50 per cent of respondents to the global survey, with optimism and indifference least reported. 'Climate anxiety and dissatisfaction with government responses are widespread in children and young people in countries across the world and impact their daily functioning,' the study found. 'A perceived failure by governments to respond to the climate crisis is associated with increased distress.'

A separate study of 530 young people aged 16–24 years old in the United Kingdom conducted by Lawrence and academic colleagues from the Imperial College and Mental Health Innovations, between August and October 2020, found feelings of anger, outrage, concern, disgust, shame, guilt and disappointment were more prominent than for Covid-19. The findings also suggest that moderate climate distress in 'non-anxious' individuals may suggest that climate concerns are a common experience rather than a sign of underlying mental health issues. This distinction between eco-anxiety and other forms of mental illness was backed by the findings of a team of psychology academics at Canberra University, the Australian National University and the University of Wellington in New Zealand. The team developed and tested a thirteen-item assessment scale that captured four dimensions of eco-anxiety: affective symptoms (such as feeling nervous, anxious or on-edge), rumination (where you can't stop thinking about climate and other environmental problems), behavioural symptoms such as difficulty sleeping or enjoying social events, and anxiety about one's negative impacts on the planet. The study demonstrated that these dimensions were distinct from run-of-the-mill stress, anxiety and depression. So if you've ever felt like you're alone in the way you're feeling about climate change, rest assured: there are millions of folks everywhere who are freaking out just as much as you are, and it's to be expected given what's going on in the world right now.

TIME TO TRANSFORM

Friends, getting caught in cycles of despair and hopelessness is just about the last thing we need if we're going to accelerate the biggest transformation of our economy and society since the industrial revolution. We need to allow space to express how we are feeling, and work on actions we can take that help us connect with people and the issue rather than become more isolated. What did Suzanne Mungall do about her charged emotions? Regular exercise, a podcast or twenty and writing down her thoughts did help, and in time, her ruminations shifted into a single-minded decision: to do *something*. 'I'm a goal setter. So in my diary, I wrote: "I'm really cranky about this. I'm really sad about this. I'm going to do this." So that's how I got myself through—I would just go to my plan.' It feels like no time has passed and Mungall is now doing not just something, but so many things: running an environment and gardening group at a local primary school; hosting an 'Ordinary Eco-Mum' group on Facebook; kicking off a South Burnett regional climate action group; coordinating writing letters to local members of state and federal parliaments; holding lobbying meetings with South Burnett Regional Council calling for a climate emergency declaration to be made; and convening group and individual conversations as a facilitator with Climate For Change, an organisation we'll learn more about later. There are more plans in the pipeline.

This is a stunning transformation, more so because it happened so quickly, in just under a year, but Mungall is matter-of-fact about the whole thing. 'I'm very talkative as you can tell, and I just go "blah blah" with everybody, so that's the sort of approach I try to take, to bring a little bit of joy.' These days Mungall laughingly describes herself as a greenie, hilarious because the only ones she'd ever encountered before were 'in the shops with the crystals'. Bringing the same level of energy and enthusiasm to her climate mission is something that's seen Mungall's mental health improve, and her appetite for life shines through. Driving every action she takes is people, the centre of her climate mission. 'I've always loved people above everything else. So even though I'm the greenie now, it's not the earth itself that I particularly give a damn about. It's actually the people in it and the relationships we have. What would be the point of the world being amazing if I've got nobody to share it with?'

Greta Thunberg pointed her finger at Mungall, waking her the hell up, and she points at all of us too. I can't help but think of how angry Greta looked a few years ago, her fury at all of us for what we had done to her generation. It made her the target of jokes by trolling climate-deniers, fuelling the darkest corners of the internet with toxic abuse, fuelling more climate polarisation. Fast forward and I'm on the couch watching her on the stage at a post-Covid Fridays for Future rally in front of Berlin's Reichstag, surrounded by tens of thousands of schoolkids, microphone in hand, hair loose, a massive grin all over her face. The serious girl, who had taken to silence and refused her food, now looks happy and bursting with life. Is this shift because the world has so radically changed in a couple of years? Nope. In 2021 Thunberg told the press she had little hope for upcoming climate talks, but finds joy in the young people she's met from around the world. Like Greta, how can we find joy in the face of this crisis without drifting into some hiatus of hopefulness that risks fuelling false or insufficient solutions?

Cobargo's Zena Armstrong believes that local, community-led capacity that's built day in, day out, is the bedrock on which her hometown's recovery will stand, a strong foundation that means people can move forward despite the inevitable setbacks and fluctuations in energy that will crop up along the way. 'It's all got to be done well ahead of any disaster, and that actually means investing in local communities, all of those groups that are the glue of local communities,' she said. 'Governments are never going to have the deep insights or the motivation of a connected and empowered community that is exploring a shared vision of the future. It's a different mindset: we're going way beyond business-as-usual.'

Jackson wrote that perhaps we've gotten the story backward; after all, overcoming monumental challenges is when wisdom and heroism are possible. 'What if the challenge confronting us isn't our greatest threat but our greatest opportunity; not a moment of self-denial and self-destruction, but of self-creation and self-discovery?' That challenge sounds like it's one worth taking on.

TOGETHER WE CAN . . . *transform our lives*

* Some 24 per cent of Australians aged between 16 and 75 years—around 4.9 million people—are 'Alarmed' about climate change, reporting feelings of uncertainty, anxiety, frustration and grief.

* Research finds that climate change exacerbates mental distress, particularly among young people and for individuals who are not directly affected.

* We can create opportunities for our world that are beyond our wildest imagination if we connect, listen and learn from our communities.

FURTHER READING

✳ Art of Hosting and Harvesting Conversations That Matter, n.d., viewed 14 February 2022:
https://artofhosting.org

✳ Australia Institute, *Polling: Bushfire crisis and concern about climate change*, briefing paper, 23 January 2020, viewed 14 February 2022:
https://australiainstitute.org.au/report/
polling-bushfire-crisis-and-concern-about-climate-change

✳ Climate Compass Audiences Segmentation (updated study forthcoming in late 2022)

✳ Cobargo Community Bushfire Recovery Fund Inc., 2022–, viewed 14 February 2022:
https://cobargorecoveryfund.com

✳ Cobargo Wellness Group, *Ginger the Frog: Helping kids shine*, 2021–, viewed 14 February 2022:
https://gingerthefrog.com

✳ Fridays for Future, 2022–, viewed 14 February 2022:
https://fridaysforfuture.org

✳ Hickman, Caroline, Marks, Elizabeth, Pihkala, Panu, et al., 'Climate anxiety in children and young people and their beliefs about government responses to climate change: A global survey', *Lancet Planetary Health*, vol. 5, no. 12, December 2021:
www.thelancet.com/journals/lanplh/article/
PIIS2542-5196(21)00278-3/fulltext

✳ Hogg, Teaghan, Stanley, Samantha, O'Brien, Léan, et al., 'The Hogg Eco-Anxiety Scale: Development and validation of a multidimensional scale', *Global Environmental Change*, vol. 71, November 2021:
https://doi.org/10.1016/j.gloenvcha.2021.102391

✳ Lawrance, Emma, Thompson, Rhiannon, Fontana, Gianluca & Jennings, Neil, *The Impact of Climate Change on Mental Health and Emotional Wellbeing: Current evidence and implications for policy and practice*, briefing paper, Grantham Institute—Climate

Change and the Environment, Imperial College London, May 2021, viewed 14 February 2022: www.imperial.ac.uk/grantham/publications/all-publications/the-impact-of-climate-change-on-mental-health-and-emotional-wellbeing -current-evidence-and-implications-for-policy-and-practice.php

✳ Lawrance, Emma, Jennings, Neil, Kioupi, Vasiliki et al., 'Young persons' psychological responses, mental health and sense of agency for the dual challenges of climate change and a global pandemic', *The Lancet Planetary Health*, viewed 17 March 2022: https//doi.org/10.2139/ssrn.3847782

✳ Patrick, Rebecca, Garad, Rhonda, Snell, Tristan et al., 'Australians report climate change as a bigger concern than COVID-19', *Journal of Climate Change and Health*, vol. 3, August 2021: https://doi.org/10.1016/j.joclim.2021.100032

✳ Patrick, Rebecca, Snell, Tristan, Gunasiri, H. et al., 'Prevalence and determinants of mental health related to climate change in Australia', *Australian and New Zealand Journal of Psychiatry* (under review)

CHAPTER 2

GRIEF IS GOOD FOR US

'Should you fear that with this pain your heart might break, remember that the heart that breaks open can hold the whole universe. Your heart is that large. Trust it. Keep breathing.'

—JOANNA MACY AND CHRIS JOHNSTONE

Ah grief, you're a reliable companion on my climate journey, tracking me like a late afternoon shadow. You materialise when I see another frightening report or look into the eyes of my two daughters, wondering what this emerging world holds for them. It's a profound sadness that waxes and wanes, dancing with surges of positive energy that fuel a vast appetite and excitement for what we can create out of this unfolding catastrophe that threatens us. How is it that we can hold these contrasting emotions in the one human brain, sometimes in the same second? Valarie Kaur wrote that grief is the price of love: the more you love, the more you'll grieve. 'When you see that pain coming, you may want to throw up the guard rails, sound the alarm, raise the flag, but you must keep the borders of your heart porous in order to love well. It is an act of surrender.' Grieving is the touchstone experience of our humanity and intrinsic to the experience of facing climate change, so what can we learn from it?

Dr Anna Seth is a busy general practitioner and founder of Tasmania's Climate Resilience Network, which aims to foster mental wellbeing in response to climate change and other environmental threats. Seth is someone who has experienced the unimaginable. Losing her first daughter to stillbirth, she said from Hobart, or nipaluna, on a blustery winter's day, helped prime her to face the profound loss that climate change has brought to her life. 'I became very familiar with sitting with grief, really intimate with grief, and really appreciative of grief's transformational power,' Seth said. 'I think going through an experience like that kind of sweeps away a lot of the bullshit.' Seth mourned her baby in the way that made sense to her, wading out into the water on a remote beach on the southernmost tip of the island state, singing lullabies to the sea as she surrendered her daughter's ashes to the ocean. 'It's hugely meaningful and comforting to know that's where she is, both the remains of her physical body and my grief, held by the ocean,' she said. 'I would not have survived without the natural world.'

It wasn't until years later that Seth had her climate moment, when the intellectual understanding of climate change became an emotional gut-punch. 'I would always vote with climate in mind, sign a petition, but I was mostly privately active, doing my recycling and so forth,' Seth said. 'I'd seen the data and seen the graphs and thought, "yes, it's serious", but I think, like for lots of people, I turned to more immediate concerns like getting the kids to day care, paying bills and sorting out clinical issues with my patients.' That all changed when Seth attended a conference and saw what she fondly dubbed her 'oh fuck' slide. What did it show? Well, here's the title, to give you a sense of it: 'Top-level causal process diagram showing major dominions of distal, intermediate and proximate harm linking climate change and mental illness.' In other words, the dull flowchart boxes and arrows connected the dots for Seth, demonstrating how climate-fuelled impacts would drive increased pressure on the health system, disrupt social cohesion, escalate isolation and more. For Seth, this mind map of growing threats to our wellbeing was something that she simply couldn't unsee, life shifting in an instant. 'It was a penny-drop moment where I was sort of overwhelmed by the enormity of this massive threat multiplier,' Seth said. 'Also I realised climate change is very much about people, not about polar bears—not that I don't care about polar bears, but it's easy not to think about polar bears.'

After coming to her climate realisation, Seth quickly began to align her personal and professional worlds around supporting people who are facing the many and varied mental-health challenges that climate change brings. 'It's not a coincidence that I've become so interested in the mental-health impacts of climate change, including grief, because I see grief as a teacher, a healer and a friend, and not to be feared,' Seth said. Any loss, she said, often comes without choice, but the question for us is what we choose to do at this moment in history. 'If you're going to choose to love things, and hold them and stand up for them, then you're going to grieve for them. Just opening your bandwidth to the full spectrum of what love looks like is letting the grief in as well.'

A BRAVE SPACE

Why have I retreated to my bedroom with the curtains drawn at 2 p.m. on a chilly late autumn Sunday? It's got to be one of the weirder online meetings of 2020, when odd is already in plentiful supply. I'm gathering for my first Good Grief group session, a program we will experience over the winter from the safety of our flickering laptop screens. Do we really want to talk about our feelings about climate change with a bunch of strangers? (Um, no thanks, says everyone, everywhere.) Yet here I sit, tucked under the doona in the coldest, darkest room in my house, away from the kids and the quizzical looks from my bloke, who's wondering if I've gone full pandemic-powered absurd. I've dragged myself to the computer screen thanks to the powerhouse that is one of Australia's first Good Grief convenors, Liz Wade. Liz is someone who, after a planned fifteen-minute chat on the phone became more than two hours of intense conversation, left me feeling more intrigued than sceptical. I have to know: can grief really be good for us? Only one way to find out, I thought with a rumbling internal grimace as I tapped in my details and committed to Good Grief's '10 steps to personal resilience and empowerment in a chaotic climate'.

'Climate grief is like your dad dying every day,' Wade says. 'We need to build the skills, supports, structures, processes and practices that enable us to hold grief without debilitating us, so we can live our lives and still hold joy, love, compassion and everything else.' Inspired by the Alcoholics Anonymous distributed model, the Good Grief Network's ten-step program was established in 2016 by two US-based social- and climate-justice folks, Aimee Lewis Reau and LaUra Schmidt. Sessions are led by facilitators who have completed 30 hours' training in the program but who are not required to be qualified counsellors or psychologists. In this way, the program can be picked up by anyone who has a passion for helping people to recognise and process heavy emotions, so these feelings can move to meaningful actions in the world. Wade, who is studying counselling and has a teaching background, saw the program on a Facebook scrolling binge and thought instantly, 'This is *the* thing.'

'We create a container that is a brave space for sharing,' Wade says, noting that one participant emphasised the 'campfire chat' feel of the sessions—gentle encouragement to share. 'The idea is not to try to fix anyone, but to be with each other.' There's logic to the concept. Psychologist Renée Lertzman noted that global climate change communications efforts have often ignored or deprioritised the emotional and experiential dimensions of climate change. These 'affective' dimensions (the feelings associated with specific practices, actions or issues), Lertzman wrote, are 'not limited to the individual, but are shared, felt, circulated and "contagious"'. When anxieties are triggered, unconscious defence mechanisms rise to protect us, including denial, paralysis, apathy and disavowal.

'Working with and addressing affective dimensions of climate change really boils down to one simple idea: creating a safe space and allowing people to have their feelings, without judgement or fear of recrimination,' Lertzman said. Turns out sharing in safe spaces is useful for our brain too. When we give each other permission to be who we are and understand how we are feeling, Lertzman said, the nervous system calms, freeing up the brain's prefrontal cortex that is in charge of executive function. Getting all the feelings heard allows us to be 'so much more capable of solving problems, being creative, being flexible, being adaptive, being our brilliant selves,' Lertzman said.

PAUSE FOR THOUGHT

Brené Brown wrote of collective joy and pain as a sacred experience: 'The problem is that we don't show up for enough of these experiences. We feel vulnerable when we lean into that kind of shared joy and pain, and so we armour up.' Now I've tried to make meditation a habit (failed, many times) and flirted with Buddhism as a life choice, until feelings of equanimity were pushed aside by another toddler tantrum or a new year's resolution that faded with summer's last sighs. There was absolutely no reason why Good Grief would be helpful for me. Besides, I'd convinced myself I was simply 'curious about the process' and 'didn't need to sit around talking about my feelings'. How absolutely wrong I was; after my first session, I was as surprised as anyone about how much I really needed this space. My protective cover-all of activity ('don't you know how busy I am saving the world and keeping up with all the important life stuff?') was shed for two-hour sessions on those precious Sundays, and over time I found myself looking forward to the conversations, a moment of collective presence amid the Covid-inspired micro-green—growing and bagel-making frenzy we were using to distract ourselves from our fears of the future.

Each week, Liz gracefully reminded us of the purpose of the session, then carefully held the awkwardness until we tiptoed into this new, brave space to share with relative strangers. The seconds of silence felt like hours until one of us made the courageous first move, skipping the small talk and arriving at the core of our feelings, usually very quickly. Then the rest of us would leap in, one by one around the circle. We spent our time wandering together through our sense of grief and loss, our fear and anxiety, but we also allowed room for wonder too: of birds in our backyard, a tall tree, our dearest family and friends and people we had only just encountered. Over the ten steps, held weekly or fortnightly to maintain consistency, the group explored a variety of topics that ranged from the philosophical ('Honour my mortality and the mortality of all' and 'Grieve the harm I have caused') to the practical ('Practise gratitude, witness beauty and create connections' and this one, exquisite in its simple direction, 'Show up'). 'It is so important to know that allowing the grief is the door to allowing the joy, love and all the other emotions that we really need,' Wade said. 'We're actually creating a kind of group mind, coming up with something that goes beyond what each person can come to themselves, that's what happens in each session.' In turning to loss, we found new friendships, good people whom I know I can go to at any time if I need to go deep on all the climate feels.

I emerged from the program like a happy-go-lucky grief-stricken crusader, evangelising about the program and asking my friends and family: why aren't we talking about our uncomfortable climate feelings early and often? Seems there are squillions of people feeling terrible, but we are trapped underneath one taboo (talking about our feelings on climate change) that's hemmed in by another taboo (talking about our feelings at all). Climate grief is like a party that no one admits to going to, where no photos are taken, and nothing is said. Our modern, disconnected culture means it's so uncomfortable for us to pause and talk through our emotions that we have a designated 24 hours—RUOK Day—where it's legitimate to ask the one simple question! Despite all this, there are some signs of progress on recognising climate change as a newer area of challenge to our mental health and emotional wellbeing. Centres of expertise on climate psychology are growing around the world, including right here at home.

Psychologist Carol Ride welcomed me into her sprawling home, on Wurundjeri Woiwurrung Country in Melbourne's north, cloistered with crimson grapevines, turning as the autumn approached. I landed in her cosy consulting room lined with packed bookshelves and the afternoon sun blazing in. 'Once you scratch the surface, we find that people actually are feeling very,

very troubled underneath, and there's not much of an avenue to express that,' Ride said. 'I think that if we could find more ways to allow people to talk about their feelings, in a way that is comfortable for them, we'll find more and more of it.' Ride spent almost three decades working as a couples therapist—thousands of hours in this room—before climate change reared up and delivered an emotional whack by way of Tim Flannery's breakthrough book *The Weather Makers*. 'I just couldn't focus in the same way as I had before. I just felt myself distracted by this issue,' Ride said. A few months later, a group of similarly freaked-out friends huddled around her kitchen table and founded Darebin Climate Action Now. The group officially launched in 2006 and has become a hub of community action, encouraging the area's local council to take on practical solutions that often lead the nation, and sometimes the world too. Before the council elections of 2016, seven of the nine successful candidates for the Darebin Council pledged support for declaring a climate emergency, and at its first meeting, the council became the first governing body in the world to formally make the declaration. Several neighbouring councils followed closely on Darebin's heels, and these days more than 1900 governing bodies in 34 nations, including several national governments and parliaments, have done the same. Darebin City Council has also made its mark by participating in the largest emissions reduction project by local government, where councils have pooled their electricity contracts to make the switch to renewables in 2021, taking the equivalent of 90,000 cars off the road.

Ride wanted to do more to merge her professional life with her personal commitment to climate, and in 2010 worked with colleagues in Melbourne, or Naarm, to co-found Psychology for a Safe Climate. 'We actually just do a lot of supporting each other . . . we create space in our group for people to talk about how they're feeling,' Ride said. 'The hallmark of the group is that we are open with each other.' Ride credited the group's attention to emotional wellbeing as a major reason they've managed to get so much done and maintain enthusiasm for their mission. As work in this area developed, it soon dawned on Ride that closing down her practice to focus entirely on climate change made an infinite amount of sense. The more engaged she became in exploring ways to contribute to supporting people to express their feelings and connect with others, through Psychology for a Safe Climate, the more this sense of purpose helped calm her inner tumult. Like many psychologists specialising in the area, Ride takes issue with terms like eco-anxiety, because they categorise people as having a pathological emotional state. Difficult feelings that arise from considering climate change are, she said, entirely rational responses to the state of the world, and so she prefers the term climate distress. 'It's rational for people to feel worried about the future, to feel worried about how they're going to live with the uncertainty and how unpredictable their future is compared with what they thought. It's very, very understandable.'

Writing on bushfire recovery, academics Bhiamie Williamson, Jessica Weir and Vanessa Cavanagh noted that Aboriginal peoples live with a sense of perpetual grief for their communities and non-human relations that stems from the ongoing impacts of colonisation. Climate change adds to the anxiety. 'The long-term effects of colonisation have meant Aboriginal communities are (for better or worse) accustomed to living with catastrophic changes to their societies and lands, adjusting and adapting to keep functioning,' they wrote. 'Experts consider these resilience traits as integral for communities to survive and recover from natural disasters.'

So is climate grief the domain only of the privileged (and I include myself here) who are starting to clock just how much advantage they stand to lose from the impacts of climate change, despite the economic and social advantage we have? 'Resilience is not just about an individual's moral fibre; the systems and structures around individuals make it easier or more difficult for people to be resilient,' Seth said. 'Rather than coming to a workshop about climate grief, people might need a replacement for their wood heater and a well-insulated home. You've got to have a certain level of privilege to be able to explore feelings of climate grief. I get that some people see it as indulgent, but I still don't think that means that we can't talk about it.'

Ride said that, regardless of why feelings of grief and eco-anxiety are manifesting, it is happening in increasing numbers so we must work to address it, crucially with action on climate at the heart. 'Not all of us in Australia have the capacity to have security in our lives; we're not threatened with loss of income or loss of livelihood or starving or impacted by extreme weather events all the time,' she said. 'We are in a privileged position in that we are relatively safe, so we have the capacity then to focus on this issue and then be able to think about what we can do about it.'

A TIME AND A PLACE

Just before Christmas every year, I wander up a hill to a North Sydney church to a space crafted to hold grief that I am, so far, lucky enough to have been spared. This place is where people come from across the city to mourn lost children, many who have died at birth. Our choir sings for them. 'This,' Reverend Graham Long says from the pulpit, 'is the place where you can lament. It's where no one will ask you if you're "over it".' The pain in the church is so overwhelming, the air vibrates with it. Tracey, one of our longest-standing members (she's an alto stalwart and #1 lady in charge of staging), stands in the alcove beside the altar, giving us some tips to get through: in case you're wondering, it's impossible to cry and sing anything, let alone belt out some good old-fashioned gospel music. 'Stay grounded while you're singing,' she says. 'Remember where your feet are. Breathe.' Before my first time here, on Tracey's advice, I clutched a 50-cent coin tightly, forcing its angular edges to bite into my sweating palm. Getting through this service is tough every single time, but at the end of the hour or so, our eclectic band of warblers are drying our eyes and sharing a hug or three. Before long we are back to our impossibly irreverent pre-dinner wisecracks at the pub down the road. The emotional wave washes over, the immensity of the privilege we have to support the space begins to settle. It's magical.

Rev. Long is a legendary figure in Sydney, pastor at large, an emeritus preacher of sorts, of the Wayside Chapel, a Kings Cross institution where the homeless and disadvantaged gather for comfort, support, prayer and a host of life skills. Graham conducted at least one funeral a week back when he was leading Wayside full time, and has led this service every year I've come along to sing. Despite this deep professional experience with grief of all shapes and sizes, he shares his tears from the pews every year, mourning the sudden loss of his son James from a stroke at only 31 years old. Loss, Graham told me later, is an intensely personal journey but one we've all got to take one way or another.

I dodged my first opportunity to be part of this service, in my first year in the choir. I pretended I was super busy/had something else on/sorry maybe next time—but the truth is I was terrified, too scared of what it would feel like to bear witness to the pain and loss these families were gathered to sit with, too scared to share my emotions with people whom I'd never met: parents, grandparents and children. They lined the pews, many weeping, as a photograph of every lost child flashed onto screens set around the cavernous church. Over the years I began to recognise the families who returned year after year, taking their place in the pews to grieve, and I became calmer in the face of this sea of loss. Ritual, it seems, helped the congregation as much as it helped me to be there simply to witness the pain of strangers. Good Grief's Liz Wade reckons that, as many people have drifted from organised religion over the decades, we've thrown the baby out with the bathwater when it comes to rituals that can be deeply nourishing. 'Those same kinds of ritual practices don't necessarily have to be connected to a church,' she said. 'Being in nature is a valuable kind of ritual.' Social science researcher Karla McLaren wrote that ritual helps us navigate and survive the passages of our lives. 'Our unfortunate disconnection from meaningful ritual not only strips us of community and the sacred, but also of our ability to live, love, feel, and grieve fully.'

Falling apart, author and teacher Joanna Macy wrote, is essential to evolutionary and psychological transformation, as much as 'the cracking of outgrown shells'. For me, the more often I've shown up to the November service, the less reluctant I am to walk up that hill. The more able I've become to free up the fear of pain, the easier it's felt to open the door to empathy. I've become freer to smile and sing through the tears that always arrive, and the more I'm prepared for others to see me in this out-of-control emotional state, to be a little more vulnerable. I'm doing what clinical psychologist turned climate warrior and founder of The Climate Mobilization, Margaret Klein Salamon, described as working out our 'emotional muscle'. 'If we stop ourselves from feeling grief,' Salamon wrote, 'we stop ourselves from emotionally processing the reality of our loss . . . Grief ensures we don't get stuck in denial, living in the past or in fantasy versions of the present and future.' Exercising our emotional capacity, she explained, helps us grow stronger by noticing, identifying and tolerating our feelings. 'Getting comfortable identifying and sharing these kinds of feelings, and having them received non-judgmentally, prepares you to share other, more difficult feelings; it also prepares you for much more challenging work.'

SWEET RELIEF

As Lertzman suggested, we're going to have to provide emotional relief to unlock our best, brightest, most imaginative selves to create the solutions we need. So how can we get that relief? I asked Carol Ride for her expert opinion. 'We feel grief because we love something or someone; it's related to that deep emotional connection that we can have with nature, with people in our lives, with animals, with plants, with where we live,' Ride said. 'We've got to connect with our enormous capacity for love.' Psychology for a Safe Climate delivers one-off, inexpensive workshops in person or online, where people come together to talk in a structured group environment, using techniques like drawing, conversations and, where possible, visits to nature during breaks in the agenda. 'People will draw themselves in a bed with the blinds down, and the gadgets in front of them, cut off from their social life,' Ride said. 'The lid is lifted on that by talking about it with others.' Ride emphasised that it's not the facilitator who delivers the results, rather that the group does the work to provide the people with the validation we all need to hear: that we're not alone in the way we are feeling. 'Look, everyone else is feeling versions of that too and, boy, what a relief to talk about it instead of having to hide it in hyperactivity and distraction,' she said. Individual counselling can help absolutely, she said, but only when it's conducted by a psychologist who understands. Psychology for a Safe Climate is establishing a climate-aware practitioners network of qualified psychologists and therapists who can offer their skills to provide one-to-one support and group experiences to people surfing emotional waves on the state of the planet.

One of the most important things the experts agree on, whether you're doing this on your own, with a therapist or in a group, is the benefits of acknowledging your feelings. Climate scientists are doing just that: writing letters that demonstrate that the individuals at the pinnacle of scientific expertise, the very people usually confined to facts and data, are just as grief-stricken as the rest of us. Their letters—part of the Is This How You Feel? project organised by science communicator Joe Duggan (an example is given in the box 'Letter to Joe')—are a collection of grief, anger and despair, but there is optimism, hope and some excitement too. Reading through these letters shows how humans are capable of holding complex and contrasting emotions simultaneously and how we can have more agency in directing our ways of being in the world as a result of talking about these feelings. A helpful framework for acknowledging feelings, known as RAIN, came my way via Tara Brach, a clinical psychologist and Buddhist teacher whom I discovered when searching through meditation podcasts during one of those new year resolutions to be more mindful:

* R: Recognising what is happening
* A: Allowing life to be just what it is
* I: Investigating inner experience with gentle attention
* N: Nurturing ourselves in our experiences

You could answer these questions in your head or journal when charged emotions start cropping up. Give it a try!

LETTER TO JOE

Dear Joe,

How I feel changes from day to day. I used to feel a lot of anger and frustration about inaction and deliberate misinformation, but it feels that we are beyond that now. Let these people have their madness; climate change doesn't care whether they accept its existence or not.

Lately I mainly feel guilt, and grief. My first child is due in June and I am acutely aware of how selfish it is to bring a new person into a future that looks so grim. I feel guilty for doing that to her, and for partially causing that future.

Grief is present too, particularly after last summer's devastating bushfires. The world we know is disappearing. Yes, there will still be places and people, community spirit and beauty, but they will occur—are already occurring—with a different backdrop to the safe world I grew up in. That is profoundly sad.

Sometimes I feel a mad jolt of hope. My work is on historical weather, our climate of the past. I think of the extreme weather we've seen in the 19th and early 20th centuries and hope desperately that the positive trends and broken records are part of something else. But then reality hits, and I know there can only be one explanation. Climate change is real, it's bad and it's us.

And sometimes I feel a much stronger form of hope, in us as people, to harness the many opportunities to work together, to make the most of the brilliance and technology we have as a species, and steer ourselves towards a safer future. It's that hope that I will pass on to my daughter.

Sincerely,
Linden Ashcroft
Climate researcher and lecturer,
University of Melbourne

Source: www.isthishowyoufeel.com/this-is-how-scientists-feel.html#linden

From her Castlemaine home in regional Victoria, psychologist Susie Burke pointed me to acceptance and commitment therapy (ACT), an approach developed by psychologists Steven Hayes, Kirk Strosahl and Kelly Wilson in the mid-1980s. It's designed to help us build our skills in psychological flexibility, which ACT therapy expert Russell Harris defined as 'the ability to be in the present moment with full awareness and openness to our experience, and to take action guided by our values.'

> The greater our ability to be fully conscious, to be open to our experience, and to act on our values, the greater our quality of life because we can respond far more effectively to the problems and challenges life inevitably brings. Furthermore, through engaging fully in our life and allowing our values to guide us, we develop a sense of meaning and purpose, and we experience a sense of vitality.

ACT is a unique approach because it intentionally avoids the reduction of symptoms, but is proven to deliver symptom reduction as a by-product. Confused? Seth is here to help. 'Whether we reason with, fight or ignore our thoughts and feelings, they will still keep on showing up and some of them will hurt,' she wrote. ACT, she said, encourages us to make space for feelings and observe with curiosity and compassion. 'The original source material for the entirety of human experience is the living world and our relationship with it. While intellect is obviously a crucial way of interpreting this, it's only part of the spectrum of our reality. We need our hearts as much as our minds.' Along with writing about the benefits of ritual, Karla McLaren reminded us that while grief is a lengthy and profound process, if we focus attention in the body, we can gracefully move into our own grief.

'Your body is a brilliant mourner, and if you trust it, it will convey you into the river of tears and bring you back out safe again.'

So how does it work? ACT's acronym is also, handily, the process:

A = Accept your thoughts and feelings, and be present
C = Choose a valued direction
T = Take action

This book will explore many ways you can direct climate action wherever you are, so for now let's stick with how you might be more mindful in sitting with uncomfortable climate emotions, or help folks around you sit with these feelings, in the same way you sit with a friend or your faithful canine/feline friend. I think of it this way: when a difficult emotion rises up, notice it and take a breath or two, approach it with curiosity and interest so you can observe what's going on. Then take a second, a minute or more to offer a little self-compassion, a moment of comfort that you'd give in a heartbeat to a close friend or a child who was hurting. Finally make a choice to act, but critically, in line with your values, and crucially not as a simple reaction to the emotional state you're in. What action can you take right now, later today, tomorrow or next week that could make a small contribution to realising the world you want to see? This process can help you see these feelings as powerful gifts to propel you towards your unique contribution to healing the world.

Good Grief taught me many self-compassion and gratitude exercises that have been helpful for me to acknowledge my climate feels, and you can find a plethora of these on my website or in Climate Action Starts Here, at the end of this book. But what about action in line with our values; can it really offer relief from stressful emotions? The evidence suggests action, particularly with other people, can offer us solace in these challenging times. Researchers at the Imperial College of London suggested climate action may support positive mental health outcomes and reduce psychological distress and anxiety: 'Strategies that help people to engage their values and goals in "meaning-focused" activities and appraisals can induce positive emotions that aid them to constructively cope with distress and anxiety. Such "meaning-focused coping" strategies may help children and young people to constructively cope with and act on climate change.' Psychology for a Safe Climate also recommends action with others as an essential strategy for managing climate distress: Ride emphasised a key benefit of acknowledging your climate feelings and sharing these with others is 'post-traumatic growth', where people can find new parts of themselves and use them for productive and ethical purposes. 'If people can do the emotional work of actually discovering and accepting what they feel and explore what it means for them, and what this means about the future of their life, it can actually be a support to the personal crisis we're in.' Grief, perhaps, is the path we must all walk with a little more confidence and a little less fear to get ourselves moving out of this mess.

TOGETHER WE CAN . . . *understand grief*

✳ Grief, anxiety and other uncomfortable feelings about climate change are *normal* reactions to what's going on; it's not a mental illness or pathology.

✳ Humans need recognition, ritual and reassurance to ride waves of uncomfortable feelings. Support in groups is particularly helpful, as is individual support from climate-aware practitioners.

✳ Climate distress can spark growth rather than cascading cycles of despair, if you acknowledge your emotions, practise self-compassion and take action in line with your values, with people who feel the same way.

FURTHER READING

✳ Duggan, Joe, Is This How You Feel?, 2014–, viewed 14 February 2022:
www.isthishowyoufeel.com

✳ Flannery, Tim, *The Weather Makers: The history and future impact of climate change*, Text Publishing, Melbourne, 2005

✳ Good Grief Network, 2022–, viewed 14 February 2022:
www.goodgriefnetwork.org

✳ Harris, Russ, *ACT made simple: An easy-to-read primer on acceptance and commitment therapy*, Oakland, CA, New Harbinger Publications, 2009

✳ Lertzman, Renée, 'Breaking the climate fear taboo: Why feelings matter for our climate change communications', *Sightline Institute*, 12 March 2014, viewed 14 February 2022:
www.sightline.org/2014/03/12/breaking-the-climate-fear-taboo

✳ McLaren, Karla, 'Grief: The deep river of the soul', n.d., viewed 14 February 2022:
https://karlamclaren.com/grief-the-deep-river-of-the-soul

* Psychology for a Safe Climate, 2021–, viewed 14 February 2022: www.psychologyforasafeclimate.org

* Salamon, Margaret Klein, with Gage, Molly, *Facing the Climate Emergency: How to transform yourself with climate truth,* New Society Publishers, Gabriola Island, BC, Canada, 2020

* Seth, Anna, 'Acceptance and Commitment Therapy: A helpful tool for coping with climate distress', Climate Resilience Network, 22 March 2021, viewed 14 February 2022: www.climateresiliencenetwork.org/acceptance-and-commitment-therapy-a-helpful-tool-for-coping-with-climate-distress

* Williamson, Bhiamie, Weir, Jessica & Cavanagh, Vanessa, 'Strength from perpetual grief: How Aboriginal people experience the bushfire crisis', *The Conversation*, 10 January 2020, viewed 14 February 2022: https://theconversation.com/strength-from-perpetual-grief-how-aboriginal-people-experience-the-bushfire-crisis-129448

CHAPTER 3

THE TIPPING POINTS WE NEED

'Although you may have a strong opinion about it, there's no way to prove you're right until the future actually happens.'

—DR DONELLA (DANA) H. MEADOWS

If you've spent more than five seconds running over the science of climate change, you'll quickly arrive, all wild-eyed and dishevelled, at the stormy cliff face of devastation. Peering over the edge of this, you'll see runaway loss of ice sheets accelerating sea-level rise, forests and permafrost releasing their carbon stores into the atmosphere, the disabling of our oceans' circulatory system, and more. A ten-second google of 'climate change tipping points' tosses up descriptions such as 'cascades', 'domino effects' and 'irreversible', and with characters like these crowding this algorithmic horror story, it's understandable that you could feel at sea from time to time. One brisk August morning, I found myself buffeted by the news that scientists had detected warning signs that the Gulf Stream, the swift ocean current bringing warmer water from the Gulf of Mexico into the Atlantic Ocean, might collapse within a decade or two ... or several centuries from now. It's confusing and scary, because each scenario is entirely possible, and if it does occur, weather systems across continents will be severely disrupted, radically altering food systems, habitats and our lives too.

These terrifying predictions, laden with uncertainty, surge with every report released by the United Nations Intergovernmental Panel on Climate Change (IPCC) and every time an 'unprecedented' flood, fire or drought dominates the airwaves. But here's the thing: tipping points are not inherently negative, or positive either. A tipping point is a 'perturbation' that causes a qualitative change in a complex system, either its future state or its trajectory. A relatively small thing happens that triggers a shift in the dynamics of the system, which then in turn triggers self-amplifying feedback that moves the system from one state or way of operating to another. Systems thinkers are obsessed with tipping points of all kinds, spending their waking hours thinking (and nights dreaming) about them. When it comes to transformation, positive tipping points are critical: they're the levers we can pull to deliver the ultimate bang for buck for our efforts to accelerate climate solutions.

So it's time to embark on a search for signs of the tipping points we need to navigate through the turbulent times we're in. A newspaper editor once told me that if you're looking for answers, it's best to follow the money, so I hauled in Tim Buckley. As founder and Director of Climate Energy Finance, Buckley has more than a quarter of a century analysing financial markets under his belt, around 25 years more than most of us. When we talked about how climate change is turning the tide in financial circles he got increasingly excitable and a bit sweary too: he was enthused not by the possibility of what might occur, but by what's shifting now and how quickly it's happening. Investors are just as human as the rest of us, following where the winds of wealth might be blowing next, and the dollars rushing to climate action these days are a hurricane. 'We're in two races,' Buckley said. 'One is the natural scientific tipping points that are cascading effects that are cumulative—all the crap that I don't look at, because I know it's a climate emergency and I don't need negative reminders every day—but the other is the tipping points in finance and that is equally as optimistic, if not more so.' With the right financial and policy frameworks, Buckley said, we can secure the right technology breakthroughs that, combined with the necessity and urgency of action, mean global finance can turn on a dime.

Buckley regularly meets with senior executives of many of the 30 biggest non-governmental financial institutions in Australia, and the biggest problem they're facing, he said, is that they cannot deploy investment dollars fast enough in zero emissions industries. How much money are we talking about? 'There is literally an unlimited amount of money in the world for the right projects,' he said. 'The bigger the amount of money, the more the opportunity.' Buckley pointed to not one, but five tipping points in twelve months that helped turn global finance around. First there was a letter to CEOs from BlackRock's founder, CEO and chair Larry Fink. 'Larry Fink is 50 times more powerful than any prime minister of Australia and yet he's unheard of by most people,' Buckley said. BlackRock manages around US$9.5 trillion in assets, which is about nine times the size of Australia's gross domestic product. When Fink speaks, the world's investment community pays close attention. As a massive campaign targeting the behemoth ran by climate activists around the globe built momentum, Fink shifted the company's position on thermal coal and now talks of a 'seismic reallocation of capital' driven by climate risk.

That's only tipping point number one on Buckley's list. In May 2021, the inherently conservative International Energy Agency released its 1.5 degrees pathway, which set out what a net zero pledge actually means: no new coal, oil or gas projects constructed anywhere in the world. Then there's the new coalition of more than 450 banks, insurers and asset managers across 45 countries led by former Bank of England governor Mark Carney, known as the Glasgow Financial Alliance for Net Zero. The Alliance has pledged up to US$130 trillion—that's US$130,000,000,000,000!—of private capital to achieve net zero emissions by 2050. Another? In September 2020, China's President Xi Jinping declared that the country will aim for carbon emissions to peak before 2030 and to be carbon neutral before 2060. Finally there was the colossal move by Mukesh Ambani, Asia's richest person and chairman of Reliance Industries, who pledged to go net zero emissions by 2035 and set a target of investing US$10 billion in zero emissions manufacturing facilities across India in the coming three years. 'It's going to be a big fight for a decade,' Buckley said. 'But when Larry Fink is on our side and when the richest man in Asia is on our side, either they're bullshitting us or they're gonna make it happen.'

UNCERTAINTY IS USEFUL

There is something delicious about uncertainty, do you agree? You and I already know the answer: when we boil it down, we all loathe it. I like to think of myself as pretty comfortable with taking risks: I'm one of the early evening dance-floor folks at a wedding (it's a sight to see) and I'll often be first to make the decision to keep pushing uphill and clambering over rocky outcrops when my heart is beating in my throat, my hands quaking. But why do I demand my children hug me goodbye before they tumble out the door, off to just another day at school? Why do I swallow a couple of vitamin pills every morning? Why do I choose to walk carefully back down my local bush tracks, holding up the folks who are gravity-propelled, sprinting south without a care? Advocates and big industry types, who also happen to be humans with mere human brains, call for 'policy certainty', guaranteed levels of risk and predictability of outcomes too. Our decision-makers rely on complex modelling to look at how certain the good times or bad times will be, all working in the way a well-functioning democracy should, so they can make decisions that are in the best interests of the citizenry, to keep the ship sailing when the weather closes in.

The truth is, humans are inclined to take any action that will reduce risk and remove uncertainty in our lives, and if we can't, as Douglas Adams quipped in *The Hitchhiker's Guide to the Galaxy*, 'we demand rigidly defined areas of doubt and uncertainty!' There's an uncomfortable truth here that resonates when I carefully turn over climate risk in my head: it's the uncertainty around the *degree* of chaos we may come to experience that I find really unsettling. We can read dozens of reports to look at the likelihood of climate doom, and we can take scores of actions that are aimed at reducing uncertainty of the impacts coming down the line, but we can't ever know exactly when a bushfire, heatwave or flood is coming or how bad it could be for our health, homes and communities. It's confronting to consider what an uncertain future brings for our families and local environment, let alone what it could mean for the people and ecosystems of the whole planet.

What's more, humans have this incredible skill in chronically underestimating our ability to solve our greatest challenges. Too easily we get stuck in the world as it is, forgetting that we have the power, intelligence and means to create the world as it could be. Why? Because every human is an expert in what eminent psychologist and economist Daniel Kahneman described as loss aversion: our brains are oriented towards trying to avoid losses rather than achieve gains. Margie Warrell wrote of how we over-estimate the risks of potential change, catastrophising the potential consequences of our choices, and underestimate the benefits that could flow from taking those risks. We need to develop skills to hold the fear of the unknown that climate change brings us in one hand, and in the other the equally high levels of uncertainty that things could turn out okay.

Uncertainty could also give us the drive we need to meet our challenges head on. In 2016, researchers at University College London decided to dig in on how uncertainty affects us, looking particularly at the effects of ambiguity on stress levels and perform-ance. Researchers led by Archy de Berker ran an experiment to assess the effects of uncertainty on the mind and body. (If you're wondering, these days de Berker builds data-driven products that help reduce or mitigate the effects of climate change.) Participants played a computer game where they turned over rocks that could have a snake hidden underneath. If a snake was present, an elec-tric shock was delivered to the back of the left hand. Over time, participants could learn which rocks were more likely to be refuges for the snakes, but the probability of the snake's presence under the rocks kept changing, using a sophisticated computer model. As the experiment evolved, the human guinea pigs had to keep recalibrating their beliefs about how certain it was that a snake would appear where it was expected. Ouch.

The research measured self-reported levels of stress and the objective indicators like pupil dilation, how sweaty the participants' skin was, and levels of the stress hormone cortisol in their saliva. The results revealed insights that we should bear in mind when thinking about (and stressing about) uncertainty in the chaotic times we're in. The study found that participants had higher stress levels when facing uncertainty than when they were certain of a negative event occurring. If you've ever felt like no news is worse than bad news, like the minutes ticking away as you wait for the outcome of a job interview or the days waiting for your doctor to call with the results of a medical test, it makes a lot of sense. A fascinating finding in the study was on performance: participants who had the highest stress responses at the times of greatest uncertainty were *better at judging whether the snake was under the rock*. The research indicated that the stress arising from uncertainty could give us an edge in solving problems and assessing risks.

LEAPING AHEAD

RethinkX is a think-tank focused on discovering the positive tipping points we could create to solve this seemingly intractable climate dilemma. Founded by Stanford University technology expert Tony Seba, the organisation's 2021 report *Rethinking Climate Change* predicted that innovative disruptions in energy, transport and food production can directly eliminate 90 per cent of net climate pollution by 2035 by using technologies we have available right now. It sounds like pretty utopian techno-babble, but there's complex modelling backing the claims. The numbers add up to what could be our reality if we make the transformative choices now, shifting from an extraction-based, centralised system that exploits scarce resources and labour to a distributed, interconnected and networked generative system where we create what we need from what's readily available.

What strikes me the most in RethinkX's report is how we've underestimated the leaps in technology that are possible: it's like we've got a threshold in our collective consciousness we struggle to cross. In 2014, intelligent, hyper-qualified folks at the IPCC predicted that, at best, solar, wind and geothermal power combined would provide only 4 per cent of the world's energy by 2100. What happened? Exponential scaling of these technologies means that estimate is very likely to be beaten before 2030, *seven decades ahead* of the IPCC's 2014 prediction. Also in 2014, according to a report by SystemIQ on how the Paris Agreement is shaping the global economy, the International Energy Agency (IEA) predicted global average solar prices would fall to US$0.05 per kilowatt hour by 2050. It took only *six years* for prices to drop to that level.

Professor Martin Green has been instrumental in helping to tip the world's energy system to renewables, but you wouldn't know it from his unassuming, no-fuss manner. Teams he's led at the University of New South Wales—electronics, engineering, physics and chemistry experts—have held the world record for silicon solar photovoltaic cell efficiency for 30 of the last 39 years. 'You can have this great idea for an improved cell but getting it to work takes a lot of work, just trying different recipes,' Green said. 'We used to look for people to do our processing who had a flair for cooking because it was very similar.' Making a solar cell work better is a little like making the perfect sponge cake, the solar 'chefs' experimenting over and over and over again until a discovery rises. 'A lot of the technologies rely on secret steps where you stumble across something interesting,' Green said. 'It's always important to have a model of what you're doing and why you're doing it, but it might not work out as it should. So then you've got to come up with another model of what's going on and then try to follow that through.'

These days, the PERC cell they came up with is used in more than 90 per cent of the world's solar panels. That's a homegrown Australian innovation that's swept the globe, after Australian Shi Zhengrong was the first to scale up industry growth through China's manufacturing and American investment dollars. It's an impressive achievement, something we don't celebrate often enough. Folks who are now giants of the global renewables industry were PhD students back then, and motivated, Green said, by the social impact they could deliver through clean energy tech. And at China's top solar manufacturers, there are Aussie graduates sitting in the C-suite in roles of chief technology officer and higher up in the hierarchy of most companies.

Here's an example that will leave your head spinning: mining haulage trucks. Cleaning up these massive, noisy and smoking creatures was a top priority for homegrown renewable hydrogen proponent Andrew 'Twiggy' Forrest. In August 2021, the founder of iron ore giant Fortescue Metals Group became only the third person in the world (after the two of the company's engineers) to drive a hydrogen fuel cell–powered haul truck. Only *130 days* after setting the Fortescue Metals team on the mission to transform the company to green energy for its operations by 2030, Forrest was chuckling like a little kid with a new toy from behind the wheel of the monster vehicle. 'I must admit the scale of their achievement made me emotional,' Forrest told an industry pow-wow at the time. 'If we could transform how we power such huge machines like trucks, trains and ships, as well as massive heavy industry like Fortescue in such a short time, why had this not been done before?'

What gets in the way? Numbers, modelling and econo-techno-speak will only get us so far it seems, with the RethinkX report's authors arguing we can't easily grasp the future because we're stuck in our ways of thinking. When we think in a linear way we end up with bandaid solutions that patch up yesterday's problems, rather than going for transformative, disruptive changes we need for tomorrow. If we've got any hope of identifying both the scale of our challenges and the incredible opportunities that are ready to emerge, we must fundamentally shift our mindset to view the world through a lens of complexity and interconnectedness.

LIGHT-BULB MOMENT

It's early 2021, and I'm in front of a presentation by Professor Will Steffen, one of the world's most distinguished climate scientists, someone who spent his younger years climbing ice-laden mountains in New Zealand and in the Himalayas for fun. Steffen's in work mode today, dispassionately guiding us along the precarious pathways in the Climate Council's *Aim High, Go Fast* report that sets out what increasing emissions means for Australia, and what our nation must do to play our part to fix the problem. The slides carousel through a carnival of tragedy (complete with handy graphs and diagrams), presenting clear evidence that strongly suggests global average temperature rise will exceed 1.5 degrees Celsius in the 2030s, leaving critical ecosystems like the Great Barrier Reef further damaged or destroyed. There's a gritty resolve that drifts from the tiny windows into my colleagues' homes, and I feel the familiar despondency fog creeping over me as slide after slide flashes on the screen. I'm thinking about asking the scary question, but have decided to do the human thing and avoid it, but someone braver than me jumps in five minutes before the hour is up: 'What about tipping points, how much more likely are they under these scenarios?' I steel myself for the usual answers that bring the horror and sadness to rest in my chest, but Steffen's answer is surprising, resetting my brain and body in an instant. 'I am optimistic,' he says, 'because we know that society can create tipping points—that's why the work you're doing is so important.'

Steffen, an Emeritus Professor at the Australian National University, has spent his career specialising in carbon cycles, tipping points and complex system theory, collaborating with colleagues around the world to understand the ways our environment stands to shift if we don't address carbon pollution quickly enough. (Fun fact: Will was transitioned out of his first career in inorganic chemistry when his job doing X-ray crystallography was replaced by computers.) Will and I caught up as Sydney rode the second wave of the pandemic sweeping through the suburbs and tipping into Victoria, which had also just locked down for the fifth time.

'Complex system theory is that we generally, or at least in many cases, really don't live in a linear world,' Steffen said. 'We live in reasonably well-defined states, and transitions between them, rather than a nice, smooth, linear curve. That's true of a lot of the features of the Earth System geophysically: they're very stable and they seem to only move a little bit when you push them high until you get to a critical point, and then you get a rapid change to a new state, and that's because of intrinsic feedbacks in these systems.' Australia crossed a tipping point in the 2019–2020 bushfire season, Steffen stated, and as a former executive director of the International Geosphere–Biosphere Programme, he is well placed to say so. 'Normally the eastern Australian forests will burn every year to some extent, but on average only 2 per cent burns, that's a long-term average, and in a bad year 3 per cent. The 2019–2020 fires burned 21 per cent of our eastern forests: it's an order of magnitude bigger; it just jumped.' So friends, if over that horror summer you felt like we were experiencing a colossal tipping point in our natural systems driven by climate change, science has got your back.

The scientific community agrees that the closer we get to 2 degrees of warming, the more likely that the frequency and intensity of these climate-driven events will increase, but how often and by how much is another matter. 'This is a risk game,' Steffen said. 'My rule of thumb is, once you start moving above 1.5 degrees you're moving quickly into really dangerous territory, where we certainly can't guarantee we're not going to initiate a cascade of some sort.' So there's nothing but uncertainty coming our way, which we're going to have to learn how to hold with clear-eyed confidence: after all, the threats of cascading destruction don't change what we need to do, and that's to reduce pollution as fast as possible. And if we're going to do that in the time that's required to maximise our chances of staying close to 1.5 degrees of warming, we'll need to bring on climate-positive tipping points too. Technology is one of the easiest ways to grasp the potential of positive tipping points. There are dozens of examples of how our lives have changed across the last 100-odd years, from power to cars and air travel to the humble telephone. Surely we can bring on exponential rates of the change we need to address climate change.

Climate scientist and tipping points expert Professor Timothy Lenton wrote that bringing a system to a tipping point usually requires some 'forcing', which weakens feedback loops that seek to 'balance' the system, keeping the status quo in place. Take this example of electric vehicles Lenton used as a handy guide to typical bang-for-buck moments that drive transformative change (Figure 3.1).

FIGURE 3.1 ELECTRIC VEHICLE TIPPING POINTS CASCADE

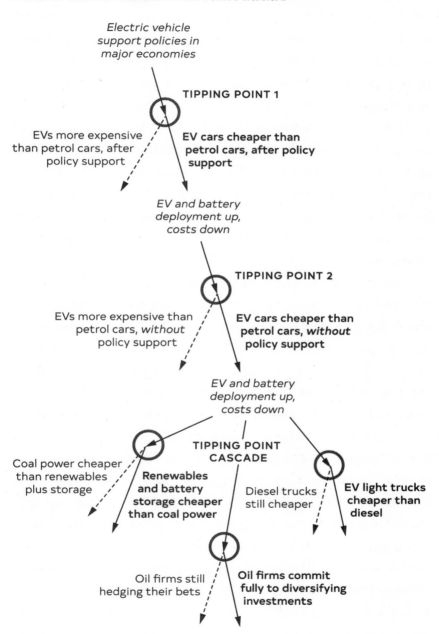

Electric vehicle support policies in major economies

TIPPING POINT 1

EVs more expensive than petrol cars, after policy support

EV cars cheaper than petrol cars, after policy support

EV and battery deployment up, costs down

TIPPING POINT 2

EVs more expensive than petrol cars, *without* policy support

EV cars cheaper than petrol cars, *without* policy support

EV and battery deployment up, costs down

TIPPING POINT CASCADE

Coal power cheaper than renewables plus storage

Renewables and battery storage cheaper than coal power

Diesel trucks still cheaper

EV light trucks cheaper than diesel

Oil firms still hedging their bets

Oil firms commit fully to diversifying investments

Source: With permission from T.M. Lenton (adapted)

Norway has an impressive record when it comes to electric vehicles (EVs), at more than 50 per cent of new car sales in 2019: according to the IEA, at that time it was more than ten times the proportion of any other country. In 2020, Norway became the first country in the world to sell more electric cars than petrol, hybrid and diesel engines combined. How did this happen? Turns out Norway used EVs as part of its push to reach 50 per cent emissions reduction by 2030, providing subsidies and other incentives. EVs in Norway sometimes cost less, but more importantly, they at the very least don't cost any more than a petrol-guzzler. Policies to bring down the price of cars created a cascade of tipping points that eventually meant that a return to the status quo was impossible.

Could something like Norway ever happen in this great southern land hamstrung by a polarised climate debate? Is there just one example that can show how we can prompt the tipping points and cascades we need? Friends, may I present the humble feed-in tariff. The name was obviously dreamed up by a policy wonk, but the idea was refreshingly simple. Governments developed programs that guaranteed a price for generating electricity from sun-power on your rooftop, with the price set high enough to ensure installing going solar was going to deliver a strong return.

These programs stayed in place until solar came so far down the cost curve that the incentives were largely no longer needed. Indeed, the success of these programs across Australian states and territories means that Australia has the honour of the highest uptake of rooftop solar around the world, with more than 3 million systems installed on Australian homes (Figure 3.2). That's not where it ends either: similar feed-in tariff-style programs called 'reverse auctions' (again I ask, who names these things?!) have seen large-scale wind and solar developments race through our

FIGURE 3.2 AUSTRALIA SOLAR PV INSTALLATIONS SINCE APRIL 2001: TOTAL CAPACITY (KW)

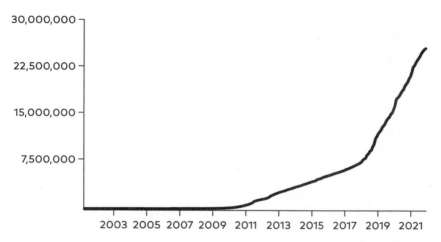

Source: Adapted from Australian Photovoltaic Institute, market analysis, 31 December 2021, https://pv-map.apvi.org.au/analyses

energy system at a pace similar to the introduction of the iPhone. If you've already gone solar, or you're buying renewable-generated electricity, be proud: you're creating the conditions for this massive tipping point in our energy system, undermining profits for coal power, Australia's number-one source of carbon pollution. We're helping to put Norway in the shade.

How did we reach this momentous tipping point in Australia's electricity system? What conditions created the cascade of solar and wind deployment that's produced thousands of jobs and started clearing up our carbon-intensive energy sector? Let's head to where it all began, in the very location where we usually hear toxic climate fearmongering masquerading as debate: our nation's capital, Canberra. Back in 2010, Phoebe Howe was only 24 when she brought a community to tip the Australian Capital Territory (ACT)'s ambition on climate action sky-high.

'I never really thought of myself as a community campaigner or activist—I was just a really ordinary kid growing up in Canberra: you know, my mum's a nurse and my dad worked in the public service,' she said. Howe spent her childhood watching David Attenborough's nature docos on TV and running around the bush, but her climate action moment kicked off when she had the opportunity through her university degree to head over to the Copenhagen climate change talks in 2009. Yes, the meeting that was called by world leaders to once and for all solve the 'greatest scientific, economic and moral challenges of our time'. Although Copenhagen was characterised by then Prime Minister Kevin Rudd and so many others as a disaster for global efforts to rid the planet of carbon pollution, it was watching how the individual advocates were making a difference around the negotiating table that sparked something for Howe. It was a tiny notion that perhaps she could do something, however small and local it might be, to make a difference.

Hitting the tarmac at home, Howe soon discovered that the ACT government was considering a special inquiry's recommendations to decide how high its level of climate ambition should go. 'A friend of mine said, "Why don't we just pressure the government to accept the upper end of the recommendation, and say 'yes we can be ambitious; we want you to be ambitious'?" And I was like, great, sounds good, let's give it a try.' The Canberra Loves 40% campaign was born a few weeks later, with Howe working in a team of local folks and organisations, advocacy newbies and stalwarts alike.

A few shifting system conditions helped make space for Canberra Loves 40% to create the conditions for more ambition. Local politics had changed: the Greens held the balance of power in the ACT parliament and a parliamentary agreement was made between the parties specifying a range of emissions reductions, renewable energy and energy efficiency measures, including legislating an emissions reduction target for the territory. Just as significant was the fact that the nation had just experienced the devastating consequences of the Millennium Drought, enormously challenging because Canberra is dry at the best of times. National debate on climate action was at a high point, with Prime Minister Rudd pledging to endorse the Kyoto Protocol. What's more, the cost of renewables was starting to rapidly decline, thanks to big US investment dollars helping China to move into the mass production of solar photovoltaics on an unprecedented scale and a growing market for wind power in Europe. The national Renewable Energy Target, introduced by the previous national Howard government, was already helping bring costs down, and solar feed-in tariff schemes were being created by state after state, with more and more systems glittering from rooftops across the suburbs. The conditions, as we say in campaigning circles, were ripening.

Canberra Loves 40%, with its love heart–stamped T-shirts and cute signs, was only the beginning. It was only a few weeks before Howe was meeting with local politicians from across the political spectrum, holding press conferences and helping connect the community with a positive vision of what the future could look like with an ambitious emissions target. Within six months, the campaign was at the pub celebrating the government's agreement to adopt the top level of the community's shared ambition, a pledge that would later be made law in Canberra's Assembly. Oh and if you're wondering, the vote was unanimous, with all parties endorsing the plan in a unique and welcome demonstration of political harmony.

Former ACT Environment, Climate Change and Energy Minister Simon Corbell was charged with making the case for the target, and he cited the wave of community support that inundated the ACT government as fundamental to seeing the legislation pass at the highest level of emissions reduction recommended. 'It convinced me about the level of community ambition,' said Corbell, who these days leads the Clean Energy Investor Group. 'It's one thing to say there's a couple of people who think this is a good idea, compared to something that can appeal to quite a broad cross-section of the community, as well as it being actually the right thing to do. Once I was convinced, I was able to convince my colleagues, eventually, that this was the right position for us to adopt.' Corbell then moved to the trickier task: implementing the territory's 40 per cent emissions reduction target (if you're wondering, it was set based on 1990 emissions levels), the first of its kind in Australia, and certainly the most ambitious at the time. After engaging some impressive advisers and creative policy experts, Corbell's team created the first reverse auction policy for large-scale renewables, which has a long-winded technical explanation but it's basically a fancy way of guaranteeing a revenue for clean energy projects: a big feed-in tariff, the very same idea that has helped clad Aussie rooftops in millions of solar panels.

A high point of rolling out the program was a short decade ago, when Canberra opened the Royalla Solar Farm, something Corbell also credits with creating a cascade of interest in the program. 'That was a real shift because, all of a sudden, everyone in the industry and everyone in the community went: "Oh these guys are serious." We had this solar farm that was on the main highway between Canberra and Cooma and the snowfields. The community realised it is a real thing; it's not just some notional political thing that I'm trying to talk about. It was real, and the industry realised it was real.' The first wind energy auction followed, and the ACT achieved 100 per cent renewable electricity in 2019, steadily building the program amid the maelstrom of the carbon tax wars taking place in the federal parliament, a couple of clicks down the road. Similar 'feed-in tariff'-type policies for large renewable projects have since been adopted by governments of all persuasions in Victoria, Queensland and significantly in New South Wales, where in 2020 the state government pledged to get a colossal 12 gigawatts of clean energy built in the state that will certainly accelerate the closure of polluting coal-fired power stations. Driving renewables into the system creates the conditions that are causing coal plants to quickly become unprofitable: in early 2022 energy company AGL announced earlier closure dates for its Bayswater and Loy Yang A power stations and Origin Energy announced Eraring, Australia's biggest coal plant by capacity, would be closed by mid-2025. With more renewables in the system, the cost of electricity trends down over time, which has seen giant businesses including Woolworths, Coles, Unilever and Coca-Cola Amatil say yes to calls from Greenpeace Australia-Pacific and RE100 to commit to 100 per cent renewable-generated electricity before 2030.

I asked Howe how she felt about playing a pivotal role in helping to create this cascade of tipping points in Australia's clean energy revolution. 'I don't want to overstate it, because I was one person and we were one group among many that have come before us, but yeah, if we hadn't been there, maybe we wouldn't have had as ambitious policies as we would have liked,' she said with the typical humility I hear from passionate people who are out there every day, getting things done on climate change. 'If it wasn't as ambitious as it turned out to be, maybe we wouldn't have had that creative spark required to build this new renewable energy acquisition policy that's now being rolled out across our country and helping us meet our federal emissions targets, when we have a federal government that's pretty unwilling to do very much at all. It's coming up from the bottom, and that's pretty exciting to see.'

DOMINO DYNAMICS

Here's the thing about tipping points that will break your brain: it is almost impossible to predict when, where or how they might happen. 'It will often come when you don't expect it, a trigger that you never would have foreseen,' Steffen said. 'You look back and you can write a beautiful paper explaining how it happened, but you can't do it ahead of time.' What we can observe, he said, are the conditions in complex but stable systems that are shifting, the environmental, social, political—and I reckon emotional too—factors that enable the system to tip. Steffen reflected that only ten years ago it felt like we'd need to make real sacrifices to get emissions under control. 'Now, this new economy is emerging so fast, and it has so many upsides, that what we're really fighting now, I think, is our vested interests and old political ideologies that are getting eroded pretty quickly.'

Timothy Lenton's work at the University of Exeter has helped to identify positive tipping points in agriculture and ecosystem regeneration, politics and public opinion that could unlock the glass-half-empty view that there's nothing to be done. But the orientation of Lenton's recent work speaks to the mindset shift we must make, individually, locally and globally, if we're going to create the world that allows us to thrive. 'We often feel disempowered as individuals when facing something as enormous as climate change,' Lenton said. 'The world is being fundamentally transformed in the history of our planet, so why not individual humans coming together and being able to identify and trigger positive tipping points to create this transformative change to sustainability?' Indeed, why not?

'Being less surprised by complex systems is mainly a matter of learning to expect, appreciate, and use the world's complexity,' wrote systems thinker Donella (Dana) H. Meadows. In other words, we should expect the unexpected, and take an approach of influencing the conditions that may cause a tipping point to occur. We're going to need grit, persistence and have a willingness to be flexible about the pathways we can see and then wander down. Most importantly we're going to need a mindset shift so we are able to welcome the possibility of the future that uncertainty creates. That's a lot to hold at once, but as Steffen said, if we cut this issue down into something we can look at right where we are, and use the best of our ability and the resources we have to offer, there might be tipping points we can help to spark.

TOGETHER WE CAN... *build momentum*

* Our brains try to avoid losses rather than achieve gains, and we overestimate the risks of potential change, catastrophising the potential consequences of our choices and underestimating the benefits that could flow from taking those risks.

* Uncertainty stresses us out, but the stress uncertainty brings could give us an edge in our performance. We need to hold uncertainty while also taking action so we can start to create the conditions for tipping points to occur.

* The tipping points we need to solve climate change are happening everywhere, and sweeping changes to global finance and Australia's energy system are only two (massive) examples.

* Look around and you might just see the opportunity to add momentum to conditions that deliver a tipping point of regional, national or even global proportions where you live, work or play.

FURTHER READING

✳ Anon., 'BlackRock invests in climate destruction', BlackRocksBigProblem, 2021, viewed 18 February 2022: https://blackrocksbigproblem.com

✳ Arbib, James, Dorr, Adam & Seba, Tony, *Rethinking Climate Change: How humanity can choose to reduce emissions 90% by 2035 through the disruption of energy, transportation and food with existing technologies*, report, RethinkX, August 2021, viewed 18 February 2022: www.rethinkx.com/climate-implications#climate-download

✳ Climate Council of Australia, *Aim High, Go Fast: Why emissions need to plummet this decade*, report, 2021, viewed 18 February 2022: www.climatecouncil.org.au/wp-content/uploads/2021/04/aim-high-go-fast-why-emissions-must-plummet-climate-council-report-210421.pdf

✳ Sharpe, Simon & Lenton, Timothy M., 'Upward-scaling tipping cascades to meet climate goals: Plausible grounds for hope, 2021', *Climate Policy*, vol. 21, no. 4, 2021, pp. 421–33: https://doi.org/10.1080/14693062.2020.1870097

✳ Turner, Meldrum & Oppenheim J., 'The Paris Effect: How the climate agreement is reshaping the global economy', 2020, viewed 22 March 2022: www.systemiq.earth/paris-effect/

✳ University of Exeter, 'Our positive tipping points are bringing change to the climate crisis', n.d., viewed 18 February 2022: www.exeter.ac.uk/research/tippingpoints

✳ RE100, 2022–, viewed 18 February 2022: www.there100.org

✳ Greenpeace, REenergise, 2022, viewed 18 February 2022: https://reenergise.org

CHAPTER 4

WE'RE CRAVING CONNECTION

'What you seek is seeking you.'

—RUMI

If you've ever felt a little isolated, take a trip to Broken Hill. After a week of winding our way to Wilyakali Country in far western New South Wales, camping along the way, saltbush and tiny towns skating past for hundreds of kilometres, we arrive blinking in the afternoon glare. In the birthplace of Australia's global giant, BHP, we wander down wide streets named for the elements that forged the outback city's recent story, a lesson in extraction on every corner. Silver, lead and zinc ores are mined here, from the world's richest deposits. It's the kind of place where a jackhammer comes in handy to dig a hole for a new tree in the parched earth, as does a dose of hope that the young roots won't cook as the seasons shift, when the soil becomes more a baker's oven than a haven for new growth. A coat of fine red earth regularly paints the buildings, cars and, if you're unlucky, the week's washing too: desert conditions boast an average annual rainfall of around 250 millimetres, less than one-fifth Sydney's 1200 millimetres. If you're like me and more at home in temperate, coastal rainforest with a good dose of summer humidity, Broken Hill pulls at your skin and itches your eyes that squint through the days under vast, unyielding skies: a tussle with the climate is a condition of entry to this place.

Back in the 1930s the woody mulga scrub that was the defining feature of the landscape was long gone, cut to build homes and fences, and used as fuel for the mines' steam engines; grazing caused more damage. The landscape took revenge, with dust storms regularly sweeping through the streets and huge sand drifts dumping at the town's edges, overwhelming fences and forcing people to leave their homes. What was to be done? Self-taught amateur botanist, Albert Morris, an assayer (that's someone who tests ores, metals and minerals to determine their quality for mining), embarked on an ambitious plan in partnership with the Zinc Corporation to use nature to build an effective shield for the silver city. Morris, who helped establish the Barrier Field Naturalists' Club in 1920, planted a small reserve in 1936 that was so successful it set the path for regenerating Broken Hill. New reserves were fenced off to the north, west and south and native vegetation planted out, healing the land. It was quite the trans-formation: these days the 'greening of the hill' is widely known as the first example of successful modern bush regeneration in Australia, inspiring conservation movements across Australia only a few decades later.

As Australia knows all too well, digging for wealth doesn't last, and as mining emptied out Broken Hill's land, the peak of prosperity fell away too, people and services drifting to other places to make a buck. What was left behind has been described by researcher Dr Jenny Onyx as profound divisions along lines of age, gender, Aboriginality and how long people stay in the area when they move here. Divisions like this disrupt social capital, described by Andrew Leigh and Nick Terrell in their excellent read *Reconnected* as the ties that bind us together in society, the trust we have for each other. In Broken Hill social capital has been in short supply of late and the impacts of a warming planet are intensifying, which is why a group of local land carers have an ambitious vision to see Broken Hill flourish again, just like Albert Morris did.

The president of Landcare Broken Hill, Simon Molesworth, has a generational pastoralist history, but in kinder locations; nevertheless, he moved his family back to this harsh outback region from Melbourne twenty-odd years ago, just as the Millennium Drought took hold, gripping the country for more than a decade. The punishing conditions killed dozens of 1000-year-old red gums guarding Willa Willyong Creek that snakes its way through the family farm. The mallee, a species better known for the steel-like strength of its wood, was not tough enough: around 80 per cent of the mallee on large areas of the Molesworth property withered and died, many close to 100 years old. Looking back at the effects of climate change creeping across the property's 4000 hectares, and seeing horticultural and other environmentally useful courses dropped first by the local TAFE then later by the only other tertiary training college in town and, considering the social disadvantage that sits uneasily with the town's wealthy history, Molesworth wondered: could Landcare make a difference? The idea germinated quickly and after a couple of years and in the midst of pandemic restrictions, the group was reinvigorated, with 350 people involved in 48-plus concurrent projects and more than

30 partners across community organisations, industry and government, the initiative winning a New South Wales Keep Australia Beautiful award for community spirit and inclusion along the way. 'Right from the Scouts and the Guides through to pensioners on walking frames, we try to find something for everyone,' he said. When we spoke, there were 2000 plants ready to go into the ground, all propagated by local volunteers, with many heading to Mutawintji National Park in a partnership arrangement with the Mutawintji Aboriginal Board of Management, assisting their Country Repair project.

Molesworth, an environmental lawyer and retired judge, doesn't have any plans to slow down: plans for a world-class sustainability hub are gaining momentum, part of Greening the Hill, Mk 2, a community-wide initiative for all to 'renew and re-green your home, your city and your life'. 'What I think we need in our Australian communities is a one-stop shop, so you can actually go to somewhere, and learn all about waste management, energy utilisation, soil and plant management, citizen science and natural history—an environmental expo site to visit to find inspiration and to learn,' Molesworth said, his eyes dancing in the same way I imagine Albert Morris's did back in the day. The hub, to be built on 3500 square metres of disused earth next to Jubilee Oval, will house a self-sufficiency education centre, community garden, bush tucker cafe, children's nature play area and a commercial-scale native plant nursery, the last to be underpinned by a seed production area to be established at the local racecourse. All elements of the development aim to create community engagement with whoever may benefit: the PCYC, service clubs, correctional services including the gaol, and First Nations–led organisations across the region. 'You know you don't have to reinvent the wheel: all this has been done somewhere in Australia or overseas before,'

Molesworth said, full of an enthusiasm for his adopted hometown that gently nudged my city-centric scepticism aside.

Broken Hill Landcare is one of 6000 groups across Australia, a network of Landcare, Coastcare, Bushcare, Dunecare, Rivercare, 'Friends of' and other community groups that involve 140,000 people. Landcare members aren't seeking national headlines; instead they are working away in the background removing weeds, healing eroded riverbanks, planting habitat corridors, building food and farming resilience programs and passing on knowledge about the environment that is specific to the local area. I asked Landcare Australia's CEO, Dr Shane Norrish, why Landcare continues to thrive, 30-odd years since an alliance of government, environmental and farming leaders came together. 'It's the fact that people have come together and remain together for the same reasons that they did when the whole thing kicked off,' Shane said. 'People were basically frustrated by the lack of attention, and the ongoing degradation in their local area, and that's what I love about it: it's people getting off their backside and responding, mobilising and making things happen.'

Landcare groups are renowned for making a difference to the local environment, but what is less recognised is how this thriving network is good medicine for people too. A 2020 study of the Landcare community found close to half of the 1000 people surveyed reported improvements in their mental wellbeing, through the simple process of becoming connected to people, communities and the environment. Of those surveyed, 90 per cent felt more connected to people; 86 per cent to their community; and 93 per cent felt more connected to the environment. The study—conducted by KPMG as the Black Summer Fires dwindled and the pandemic picked up—found that 46 per cent of people surveyed reported an improvement in mental resilience and ability to manage challenges as a result of their involvement in Landcare. For folks who reported being involved for as little as four hours or less each month, 43 per cent still reported an improvement in their mental wellbeing (see Figure 4.1). One fascinating result of the study was how folks in major cities had higher benefits than those in regional areas: 59 per cent in the major metro areas noted an improvement in their mental wellbeing through their involvement in Landcare, compared to 47 per cent in the regions.

FIGURE 4.1 IMPROVEMENT IN MENTAL WELLBEING AFTER INVOLVEMENT IN LANDCARE

Hours per month

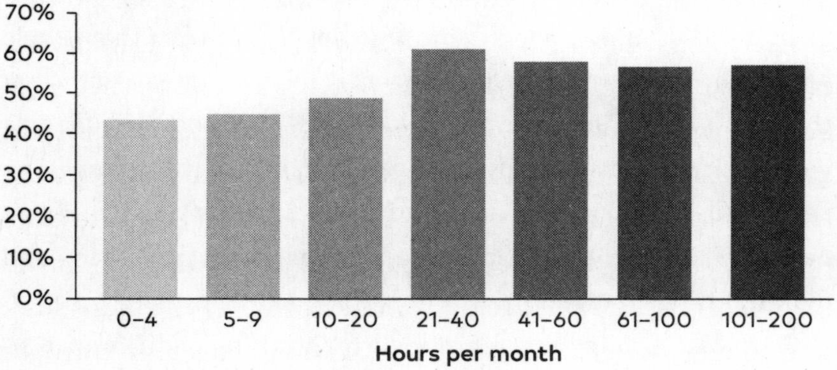

Source: Adapted from *Building Resilience in Local Communities: The wellbeing benefits from participating in Landcare*, KPMG, 2021.

ALONE WITH YOU

These days it feels like we are at the pulsing heart of a hyper-connected world, always on, linked in to our communities with the pinging and buzzing sending techno-ripples through our Saturday afternoons, interrupting breakfast, lunch and family dinners, holding a good night's sleep at arm's length. Connectivity defines the modern world and so it defines us, with a 2021 report finding

Australians spend more than 40 per cent of our waking hours online, an average of 6 hours and 13 minutes a day, an increase of 10 per cent in only twelve months. The report found one-third of that time is spent on social media, with YouTube, Facebook and Instagram the most popular platforms. Our connections are essential to life, but the flood of information through their gateways is what author Carol Sanford noted as the definition of a pollutant: a resource that's entering a system faster than it can be absorbed. I don't want to think about how many hours that adds up to over my lifetime, let alone what it means for my teen daughters.

There's a theory to explain the phenomenon of the growth of networks: Metcalfe's Law. Coined by Robert M. Metcalfe, co-inventor of Ethernet (the technology standard that enabled computer networks, better known to me as that blue cable that you plug into your machine when the wi-fi isn't working), the law basically means a network is more valuable if there are more people you can contact, or the more web pages that are linked together. Life, my friends, is complex in its connections and, according to Metcalfe's Law, the connections equate to the *value* of the network (see Figure 4.2).

I have a long list of connections across all the platforms that my age and stage gravitate towards (yep, I'm of the era that's thoroughly underwhelmed by the prospect of sending hundreds of 'snaps' to 'chat' every single day). I post, like, share and message several times a day. I enjoy staying 'in touch', and it does feel warm and fuzzy when a bunch of happy birthday messages arrive in your feed from people you haven't seen since two decades before social media arrived in the world . . . but there's something about it that doesn't sit comfortably. It feels like a parallel universe, where 'acquaintance' is the defining characteristic, not connectedness

FIGURE 4.2 METCALFE'S LAW: HOW THE VALUE OF A NETWORK GROWS

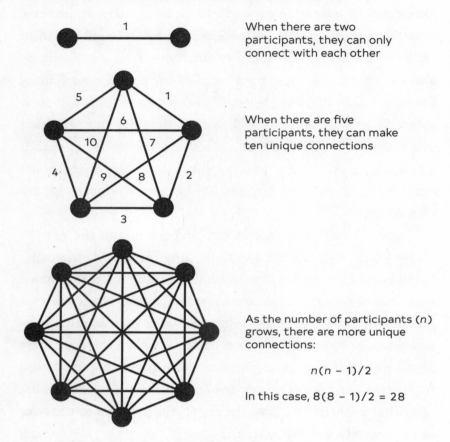

When there are two participants, they can only connect with each other

When there are five participants, they can make ten unique connections

As the number of participants (*n*) grows, there are more unique connections:

$$n(n-1)/2$$

In this case, $8(8-1)/2 = 28$

Source: www.reliantsproject.com/2020/06/14/
concept-8-metcalfes-law-and-network-effects

of any depth. Have you ever bumped into an online connection in real life, and had a conversation that felt a little emptier than you expected, a little awkward and distant? Ever encountered someone at a party when you realise you've met them before, but it's 'only through Facebook'? It's moments like these when the shallowness of digital connection is exposed, when we get a hint of the reality that the precious moments we post are just algorithmic 'content' flows that keep us scrolling, scrolling, consuming and scrolling some more. Are our connections of any 'value'? The box 'Loneliness: the numbers' offers some clues.

LONELINESS: THE NUMBERS

* 44% of Australians regularly feel lonely
* 54% of Gen Z and 51% of Millennials say they feel lonely either often, always or some of the time, which is higher than all other generations
* 48% of Australian parents with children at home under eighteen years say they feel lonely either often, always or some of the time
* 61% say that when they feel lonely they do not talk to others about it
* 41% say they are worried others will judge them if they say they are lonely

Source: Telstra, Talking Loneliness, 2021

Connectivity has grown a bunch, but at the same time we're seeing levels of loneliness swelling too. One in four Australians reported problematic levels of loneliness well before the pandemic locked us inside our homes, exacerbating our isolation. The 2018 *Australian Loneliness Report* produced by the Australian Psychological Society and Swinburne University based on a survey of 1678 people, found close to 30 per cent of Australians rarely or never feel part of a group of friends, one in five rarely or never feel close to people or feel they have someone to talk to. Nearly one-quarter say they can't find companionship when they want it. What's more, Australians rarely look to their community for assistance. According to the 2018 survey, one-third of people are disconnected from those geographically closest, with this group reporting they have no neighbours they see or hear from on a monthly basis: nearly half, 47 per cent of respondents, said they have no neighbours they can call on for help. The 2020 *Ending Loneliness Together in Australia* white paper noted single parents, people with a disability, carers, people from low socio-economic backgrounds, people with an immigrant or non-English speaking background and people living alone are more vulnerable to feeling problematic levels of loneliness. Even those whom we tend to think of as social butterflies—those aged between 18 and 25— are actually those who feel loneliest, as Figure 4.3 shows. In case you're wondering, we're lonely in the workplace too: around 37 per cent of Australian workers feel lonely among their colleagues.

Loneliness is like an iceberg, explained US social neuroscientist and loneliness expert John Cacioppo: we are conscious of the surface but it goes so deep we really can't recognise the impact it can have on our lives. Cacioppo's research revealed that loneliness is destructive if it is long term, hampering our ability to think

FIGURE 4.3 LONELINESS ACROSS THE LIFESPAN

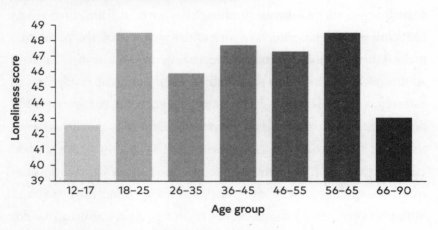

Source: *Ending Loneliness Together in Australia White Paper,* 2020

and sleep, damaging our willpower and our immune systems. Loneliness leads to poorer psychological health, lower energy levels, feeling less able to cope with problems and social situations, greater risk of social isolation and a greater tendency to suppress emotions. The physical impacts are well documented too: the *Ending Loneliness* report found loneliness is associated with a 26 per cent greater risk of premature mortality, and having poor social relationships (loneliness and social isolation) is associated with a 29 per cent increase in the incidence of coronary heart disease and a 32 per cent increase in the risk of stroke. Cacioppo also noted an important distinction: if the state is short-lived, he said, loneliness can be positive and necessary, because it highlights the need for social connections. Perhaps those days, weeks and months of pandemic lockdowns will be the teacher of our times, a society-level reminder of just how much we are craving deeper connection in our lives, and why we must prioritise creating and maintaining quality relationships if we are going to be balanced, healthy humans.

DEEP TIES

Wiradjuri author and ancient society historian Nola Turner-Jensen holds connection to place so deeply, she described Country as being fused to her genes. More than 60,000 years of continuing society in Australia's second-largest language group means the animals and plants that surround her are ancestors whom she greets with the same familiarity as of an uncle, sister or close friend. 'In our culture, you can't fix one problem without fixing all that's in relationship to it, that's the very basis of our thinking and our philosophies, everything's connected,' Nola said. 'Like a respected uncle says, you can't have connection without knowledge, and you can't have knowledge without connection.' Skyworld Totemic ancestors are at the heart of First Law, with groupings of animals and plants setting down the forensic details of how a life is lived in relationship to people and Country: what you can and can't eat, the species of plant and animal it's your duty to protect, whom you may marry, who your family is. Signposts set in the local environment are frequent reminders of Law passed down over generations. 'These visual cues are everywhere, from the tiniest little insect to the biggest kangaroo, every one of those living beings around you in your local area,' Nola said. 'So when you get up in the morning, and you hear the kookaburra, that is an ancestor, or you saw a fish jump out of the water, an ancestor. And some of them are very special to your family.'

'Durru-ga-rra Mayinybang Gunhi' (pronounced *doo roo garra May yin bung gore knee*) means 'what Country do you belong to', and it is a powerful question, one made more powerful when Nola explained that 'Gunhi' means 'mother' in Wiradjuri. Nola gently guided my colonial brain through how totemic ancestry is set in Wiradjuri Country, which stretches across New South Wales from the Blue Mountains in the east to Hay in the west, north to Nyngan and south to Albury. Coincidentally it's the Country I was born on, thanks to the accident of my father's job at the now-closed coal-fired power station named Wallerawang, which in Wiradjuri means 'a rocky waterhole where two creeks meet' (Walar means waterhole in the rocks; Wilawang means junction of two creeks). Skyworld Totemic ancestry is as complicated as getting my head around a legal case in the Westminster tradition, and Nola's life purpose is to restore Ancient Established Society for all Wiradjuri who have had it dismantled through horrific legacies of colonisation: forced family separation, intentional destruction of language and stolen land. 'If we understand the standards, the rules, the law for our society and how to live, if we start teaching that again, we will be restored,' she said. 'In a Skyworld society, who you are as an individual is not important; who you are in a kinship relationship to everything above and below ground in your local living landscape is,' Nola explained. 'Intimate knowledge, observation, songs, deep listening and ceremonies all help reveal your relationships to birds, to animals, to place and to members of your Kinship.'

As winter started its tug of war with spring, balmy days switching to chilly gusty evenings on Dharawal Country, just south of Sydney, the king parrots, rosellas and rainbow lorikeets returned for the season, searching for small gifts of seed from our back verandah. If they were lucky they got a small handful before the white cockatoos swooped in like a rowdy mob of flying footy players on a Friday night pub-crawl. Long bouts of pandemic lockdowns, with so many days closed in from life as usual, opened up my eyes, my ears and my heart to the always vibrant, sometimes demanding, visitors, who at times wandered into the kitchen seeking us out. I recognised their calls and sometimes joined them to chatter away about nothing in particular. When they arrived the loneliness of lockdown faded, and I couldn't have cared less about the irrationality of conversing with the birds over my cooling pot of tea: after all, I was in good company.

Nola suggested that perhaps I was a 'bird lady' (meaning I came from a bird Skyworld ancestor long ago), our conversation digging into what I most remembered about my local places as a child, what gave me joy or left me fearful. The images started with a flutter before they were off and flying: the magpies and kookaburras my mother would chat to as they came for a snack, my grandmother feeding the rainbow lorikeets, her gigantic hat covered in the feisty friends, their screeches of delight deafening as she stood under the magnolia tree holding up a tray laden with bread soaked in water and honey (terrible food for birds, I now know). My sister's pictures of birds all over her suburban apartment, my elder daughter's incredible recall for the tiniest identifying features of the birds we'd spot on local bushwalks— simpler times when she was half my height and free of teenage cynicism. The idea of connection to the birds was feeling a bit spooky and unreal, and it took substantial energy to just pause my hard-nosed brain confined by my pseudoscientific version of reality and go with Nola's flow, diving in. I've always felt overawed by the beauty of this country in the places I've been lucky enough to travel to, but I'd never really considered that this Irish-brained foreigner could be remotely allowed to become deeply connected to this land: I'm not entitled to it, my ancestors colonised it, likely causing unspeakable harm to the owners of the country who never ceded it. But Nola guided me through this intergenerational guilt and shame that limits so many from taking the leap to under-stand the unique and beautiful qualities of this precious place. If you're connected to a place, Nola instructed, you honour it by learning about the living things around you, so you might share in their strengths, and give them yours when they are in need. What a wonderful thing to be always connected to every living being in your environment, to not be separate or superior. 'We have this beautiful cyclical revival of our ancestors, who are always watching to make sure that we're doing the right thing, they're always there for us and ready to give their strengths to us if we need some,' Nola said. 'In our ancient world we are never lonely.'

Nola Turner-Jensen lamented the fear, guilt and shame she sees from so many folks with colonial roots: fear to engage, connect and learn from the Country and ancient culture that's alive in every plant, animal and person around us. 'Your perspective of knowing and our perspective I think would be rather different, but it doesn't mean that you couldn't learn,' she said. 'The capacity to learn our ways and thinking is what we've wanted from the very beginning but very few have been willing to take that up.' To me it feels like a symptom of the affliction of the world we live in now: we're lonelier, which begets being too wary of human-to-human contact, and so busy being 'connected' online that we don't get connected to place, people, culture, knowledge that's all around us.

Nola believes the lack of responsibility for people and places around us is getting in the way of the healing of Country and people that we desperately need. 'People hire other people to look after the koala and don't look after him themselves. A man up the road sleeping in his car: do you go and give him a meal or 50 bucks? No, you ring the Smith Family. I think it's allowing people not to be responsible for their fellow man or fellow living things.' I see complex causes that entwine economy, culture, history and social norms that all add up to the world we experience and the choices we make, but I get Nola's core point. Our lives are too preoccupied—for whatever reason—with being time-poor, stressed-out human 'doings', so much so that we let time as human 'beings' too often fall a long way down the priority list. Imagine what it could mean if we took it on ourselves to build deep connections with the natural world around us and the people who are part of it? As Nola said: 'When you put everybody before yourself, everybody, every living thing, how can you ever be wrong?'

CUP OF CONNECTION

Karta Pintingga, or Kangaroo Island, off South Australia's coast, is known for its stunning ecology and slow pace of life. Its community of around 4500 people is connected and cohesive. 'It's got a really warm, welcoming tight-knit community,' said Sabrina Davis, who calls the island home, a world away from her early years growing up behind the Wall in a small East German village south of Berlin. 'A lot of people have lived here all their lives: they're fourth-, fifth-, sixth-generation Islanders,' she said of her adopted community. 'They lived in camps when they first came here, and experienced the hardship of clearing land, so they're tough as nails.' Australia's third-largest island has a widely dispersed population where heading down the road to visit a friend can easily mean a 100-kilometre round trip, so locals stay connected through community sport that runs all year round. People here are hardy, self-reliant and practical.

In December 2019, Kangaroo Island began burning, and three weeks later 95 per cent of the national park, conservation park and wilderness areas were scorched. Two community members lost their lives to the flames and countless livestock, native animals and birds were destroyed. That summer was a nightmare for the Davis family. Sabrina lost her home and the family's third-generation sheep farm at Gosse, on the western tip of the island, a precious, isolated oasis surrounded by blue gum plantation forest and a few minutes away from the Flinders Chase National Park. Sabrina's was one of 87 families that lost their homes, and overnight she found herself relocated to Kingscote, 100 kilometres away from the devastated property. Everything resembling their previous life on the farm shifted in an instant. Reflecting back on that time, she remembered that despite being in the island's bustling heart, living with her children aged five and seven at the time, with the in-laws down the road, she felt unsettled, traumatised and lost. 'I just was going through what I call my zombie state. I was just functioning for the children: I was trying to still make some good memories during school holidays, I was trying to settle into a new routine, but I had to go and help my husband on the farm and travel a lot,' Sabrina said. Husband Ben spent most of his time back at the property, first living in a swag, then in a family caravan, then in a 'pod', a temporary emergency container donated by the Mindaroo Foundation; he was dealing at first with the immediate aftermath, later the rebuild. The endless to-do lists, mountains of grant applications and insurance forms, and the sheer physical work of the rebuild were taxing, and the arrival of Covid-19 in March 2020 after the fires soon halted recovery meetings and catch-ups. These conditions all conspired to prevent Sabrina and so many friends and neighbours across the community continuing their usual practices and habits of connecting.

Deciding to get to know her new community in town, Sabrina decided to reconnect with a friend or two over a cuppa, and the realisation of just how profound these experiences were to her healing journey opened a doorway to what could help her begin to heal. 'I think we were all at that point where we just really craved connection, craved a chat. I only in hindsight understood how special it is to have someone just listen to you,' she said. 'It's different when you sit with someone.' Not surprisingly, psychologists have come to similar conclusions, and offer similar advice, summarised in the box 'Tips for reconnection' to help people build their social networks.

Hoping to bring people back together, Davis started throwing these coffee conversations online, and her stories project and website—Humans of Kangaroo Island—was born. The chats started with closer friends, then quickly evolved to deep, emotional conversations with complete strangers who would lay out the big and small moments of their lives, their greatest joy and darkest fears over a cuppa. 'I don't think that I actually understood early

TIPS FOR RECONNECTION

LOVE QUALITY Quality and enjoyment matter more than the number of friends you have. Savour moments of connection, wherever you find them.

THINK POSITIVE Don't over-think your interactions or dwell on worries about how you are perceived—shift your focus to the other person or the topic of conversation.

TUNE IN Ask questions and really listen, rather than just waiting for a turn to talk, and respond warmly through your posture, facial expressions and words.

GET OFFLINE Social media can also increase disconnection; perhaps invite trusted online friends to a coffee, meeting or event to build your relationship in real life.

JOIN UP Embrace opportunities to join, volunteer, participate and help out. This connects you to other people, unites you in a shared activity, and provides an easy way to get to know people better.

REACH OUT Reach out to friends from your past. People welcome such efforts and the feeling that you care. If you plan a catch-up, revisit a place or experience where you shared happy memories.

NAMES MATTER Using someone's name when you know it demonstrates caring. Offer yours, and ask about their loved ones, pets or hobbies to show you've paid attention.

MANAGE STRESS Everybody has some social situations they dread (for me it's the first twenty minutes at a party), so use simple stress-management techniques, like breathing deeply and slowly, to help keep your stress in check.

TOLERATE DISCOMFORT It's normal to feel anxious or awkward when you're socialising. Reach out to others and your skills will improve the more you give it a go.

PRACTISE Remember that social connections are good for you and relationship skills can be learned. If you feel like you need support to build better connections skills, a psychologist can help.

Source: Adapted from the Australian Psychological Society: https://psychweek.org.au/2018/tips-to-connect-and-thrive/

on why I was doing it. It was really basic, to be honest; it was really just this need of talking to people, and my gut instinct told me that I wasn't the only one who needed connection.' Soon a growing circle of new friends and neighbours were stopping her on the street, asking when the next story would arrive. 'There were so many worries and so many thoughts and so much going on and I just wanted to feel carefree and happy again. And when I was with other people, I felt that way and it just energised me.' Sabrina, who caught the journalistic bug from her grandfather whom she followed around on his part-time sports reporting jobs, aimed for one conversation a week.

You've probably already guessed that the humble storytelling website isn't where this tale ends. Davis said the fire was one of the positive things that happened in her life, something that is difficult to grasp for the rest of us lucky enough not to have faced losing our entire home, or far worse, the lives of family or friends. The experience has been a teacher, opening Davis's eyes to what she is truly capable of, both mentally and physically, and sparked a new passion for disaster preparedness. The project's first fundraiser in November 2020 raised $60,000 for safety equipment for local farm firefighters, and Humans of Kangaroo Island's inaugural film and literature festival in the island's heartland, Parndana, was only a few weeks away when I caught up with Davis, who was back on her property in a converted shed within spitting distance of the rebuild of the family home. 'I believe that everyone has a story to tell, and everyone's story is unique and special and interesting,' she said. 'If I see someone I've interviewed on the main street or in the supermarket, I don't just walk past them. I stop and we have a genuine conversation because I know part of their life. It creates a bond between us.' Sabrina's idea has evolved into a wonderful example of how we could be better equipped to face tragedy if we let go of our mistrust and cynicism and shared our experiences and emotions more readily and invited people to become a little closer to us, so we can be more meaningfully connected.

TOGETHER WE CAN ... *reconnect with each other*

* Australians spend more than 40 per cent of our waking hours online, an average of 6 hours and 13 minutes a day, and one-third of that time is spent on social media.

* But our connectedness is skin-deep: 44 per cent of Australians regularly feel lonely and 41 per cent say they are worried others will judge them if they talk about it. Nearly half of Australians report having no neighbours they can call on for help.

* Form connections by learning about the animals, plants and people in your community: put your phone on silent and meet an acquaintance for a coffee—you never know, it could be the start of a new friendship that changes your perspective, perhaps your life.

FURTHER READING

* Australian Psychological Society & Swinburne University, *Australian Loneliness Report*, 2018, viewed 18 February 2022: https://psychweek.org.au/wp/wp-content/uploads/2018/11/Psychology-Week-2018-Australian-Loneliness-Report.pdf

* Community Research Connections, Royal Roads University, 'Broken Hill, Australia', CRC Research, n.d., viewed 18 February 2022:
www.crcresearch.org/social-capital/broken-hill-australia

✳ Davis, Sabrina, Humans of Kangaroo Island, n.d., viewed
  18 February 2022:
  https://humansofkangarooisland.com

✳ Ending Loneliness Together, *Ending Loneliness Together in
  Australia*, white paper, 2020, viewed 18 February 2022:
  https://endingloneliness.com.au/wp-content/uploads/2020/11/
  Ending-Loneliness-Together-in-Australia_Nov20.pdf

✳ Landcare Broken Hill, n.d., viewed 18 February 2022:
  www.landcarebrokenhill.com

✳ Leigh, Andrew & Terrell, Nick, *Reconnected: A community
  builder's handbook*, La Trobe University Press, Melbourne, 2020

✳ Telstra, *Talking Loneliness Report: Research into the state of
  loneliness in Australia in 2021*, 2021, viewed 18 February 2022:
  https://1u0b5867gsn1ez16a1p2vcj1-wpengine.netdna-ssl.com/
  wp-content/uploads/2021/10/Telstra-Talking-Loneliness-Report.pdf

✳ We Are Social, Kepios Pte Ltd & Hootsuite, 'Digital Australia
  2021', We Are Social, 2021, viewed 18 February 2022:
  https://wearesocial.com/au/blog/2021/01/digital-2021-australia

CHAPTER 5

FROM THE GROUND UP

'Start where you are. Use what you have.
Do what you can.'

—ARTHUR ASHE

There's a ray of sunshine breaking through weeks of dismal weather on Dharawal Country, so I grab the opportunity to wander around four and a half hectares of food forests, market gardens and farm animals. I'm nestled in a valley at Green Connect Farm, a thriving social enterprise hidden behind Illawarra suburbia in Warrawong just south of Wollongong, New South Wales. I've chucked on a hat that screams city dweller who's trying too hard, and switched my sneakers for some serious knee-high gumboots to plough through the aftermath of a few weeks' deluge. 'It's very muddy, but when you've got sun and rain it's happy days for farmers: we're in a golden period right now,' says Green Connect's general manager Kylie Flament. The fresh early summer breeze tumbles the honeybees among healing remnants of Berkeley Brush and rushes through the reeds in the creek that catch the run-off from surrounding brick-veneer blocks, nature's toolkit scrubbing it clean enough to water rainbowed rows of vegetables. It's a world away from crouching over the laptop on the swaying commuter train to Sydney: it feels exponentially healthier to be squelching through the mud, hearing the chickens clucking away and the tractors rumbling along, the UV prickling my skin for the first time in far too long.

This site is believed to be one of the largest urban permaculture farms in the world, but it's so unpretentious that locals living a couple of streets away didn't know it existed until Flament knocked on their doors to say hello. Back in the day, the site was part of a dairy farm, until the state's education department picked it up in the 1950s. A high school and primary school were soon constructed on the top of the breezy hill and the valley was left to the elements, the leftovers of modernising life taking up residence. 'When we came here in 2013, it was lantana, weeds and rubbish: people used it as a dumping ground,' said Flament. 'Now we've got a bunch of different animals—pigs, chickens, goats, alpacas, bees—and we grow more than ninety different types of fruit, vegetables and herbs.' It's a peaceful place for sure, but what's obvious is the quietly complex operation humming here: the farm supplies 200 families, plus local restaurants and cafes, with fresh, organic food every week. 'There's constantly vegetable beds being planted, growing, being harvested, and then turned over, usually by our pigs, goats and sheep.' During waves of the pandemic lockdowns the farm supplied boxes of food and essentials to hundreds of families in need, with the costs covered by local people who jumped online to 'donate a box'.

'For us fair food means it's good for the people that grow it so they have fair wages and good jobs,' Flament said. 'It's good for the people who eat it, it's healthy, it's pesticide free and it's good for the environment. So we're not polluting, we're not causing environmental damage downstream.' Farm manager Eh Moo, with his big smile and shovel in hand, is one of many who have complex stories that brought them to tend the market garden and animals. Moo lived his first ten years in a refugee camp on the Thai–Burma (Myanmar) border before arriving in Warrawong in the year 2000, landing in the primary school, then the high school over the back fence from the farm, where we can now hear the bell calling students back to class after morning break. 'After school I was doing a community service diploma at TAFE and we were required to do a work placement for our study,' Moo said. 'At the time I didn't like gardening much, but I've been working here now for five years and I love it.' Learning to garden has brought many lessons, including how a safe and healthy environment means healthy food that brings physical and emotional health too. 'We save money, we save time, we live longer and have a really happy life. It's really powerful to give back to Mother Earth.'

Green Connect's seeds were planted in Wollongong in 2011, when SCARF, a local refugee support organisation, was funded to build a social enterprise employing local refugees to conduct waste recovery at events. A couple of years later, when project funding withered, local Jess Moore was hired to close the operation, and learned from recently arrived refugee families that they were keen to earn their living on the land. Moore, who was completing a permaculture design course on Warrawong High's land at the time, saw potential over the back fence of the school's tarmac-covered basketball courts and set out to revive the project with permaculture at its centre. In under a decade, Green Connect has evolved into a thriving social enterprise with deep roots in the local community and no fewer than five business operations including landscaping services, a zero-waste events operation and an op-shop. The core goal for Green Connect is to create meaningful local jobs for high-unemployment groups, providing paid work and volunteering opportunities for former refugees and young people. 'We employ young people and former refugees, but sometimes people who don't fit in either of those categories; they just need a chance and no one's giving them one,' Flament said. 'We ask that they come to the farm two mornings a week for three weeks and do some work experience.'

Wollongong's natural beauty drew me in when my first love landed a place at the university, one of the region's largest employers these days. The sweeping Illawarra escarpment is stunning whether it's showcasing an epic sunset or cloaked in morning mist; the region's industrial centrepiece, Bluescope steelworks at Port Kembla, is visible from most of our coastal vantage points, a reminder of Wollongong's transitioning economy. The recent history of my adopted hometown is all steel-making and coalmining, but times are changing: the steelworks has 5000 employees where once there were four times that number, and only a couple of coalmines remain. A new nostalgia for coal has emerged in the last decade, with 'Coal Coast' beers for sale at the local bowling club, and business after business proudly displaying the black diamond. Wollongong City Council's 2019–2029 economic development strategy noted 6000 local jobs in manufacturing were lost between 2007 and 2018, constraining jobs growth to 0.5 per cent per annum over the last decade, falling well behind both regional New South Wales at 0.9 per cent and Greater Sydney at 2.1 per cent. Huddled between pristine beaches and subtropical rainforest escarpment, my hometown will see climate change determine our city's future, both environmentally and economically. This industrial city, like so many fossil fuel-dependent regional centres around Australia, is buffeted by global trade as much as domestic policy, where the rhetoric of transition is all opportunity, but the reality carries risks of loss and displacement.

At Green Connect I saw lessons for transition that are a lot bigger than the direct benefits the farm provides to workers, volunteers and families in the Illawarra. Permaculture is at the heart of the organisation, an approach often misunderstood by those unfamiliar with it for a way to plan your organic veggie patch or take up residence at a commune if you go full-blown 'permie'. I see it more as setting an ethical framework for creating and maintaining systems that benefit people and the planet, a way of approaching living our best lives. One of the global permaculture movement's Australian founders, the late Bill Mollison, described permaculture as a philosophy of working with rather than against nature, using protracted and thoughtful observation rather than protracted and thoughtless labour. The three guiding ethics of permaculture are *earth care*, *people care* and *fair share*, which are then further broken down into twelve principles (see Figure 5.1). Together they amount to an instruction manual for a life better lived. It feels like the right framework for transitioning a community, even a country, and it's all too possible if we consider carefully where we are and look at what's required to ensure a healthy interconnected community, economy and environment.

'You cannot take a young person or a former refugee, or anyone who's facing multiple and complex barriers to employment and fix just one element in isolation,' Flament said. 'Ditto, you cannot fix the environment and do terrible things to people or help people and do terrible things for the environment. We're all in this together; it all has to fit together. And on the ground, you just do what's right.' Green Connect is crafting a new future on its plot in Warrawong, recognising and working with the enormous complexity that's

FIGURE 5.1 PRINCIPLES OF PERMACULTURE

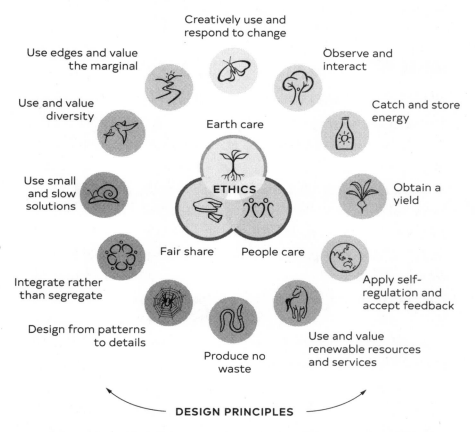

Source: Adapted from www.permacultureprinciples.com and Holmgren Design

involved in transitioning a place. It's hard, physical work each day, but you hear the laughter floating down the hillside, a cheery accompaniment to the practical attitude of the people tending the plants and animals. I tore myself away to rush back to the screen and a list of cascading deadlines, and I arrived late to the video call with half-hearted apologies and excuses, but the truth I didn't admit was that I simply didn't want to leave the farm. In a couple of short hours I saw more clearly how we could forge a new way of living by getting our hands a little closer to the ground and our hearts a little closer to each other too.

COMING HOME

I first met Yael Stone when she wandered into my office with a friend in late summer 2020, when the last of the Black Summer fires were smouldering and the news was beginning to whisper of something called a coronavirus, emerging across the planet. A quick google revealed that Yael was actually a really famous actor, who starred in *Orange is the New Black* and had a gazillion followers on social media (friends, I really am a gigantic nerd), and that we were practically neighbours, both recent arrivals to the northern Illawarra. Stone had just made headlines by pledging to dramatically reduce international travel to minimise her contribution to climate pollution, a choice that came with the inevitable consequence of relinquishing her American 'green card', the pass foreigners fight hard to secure. Giving up the privilege to work and live in the United States is a brutal decision for any performer on a soaring career trajectory to make. But reflecting on that summer, Stone said the fires were a multi-sensory experience that overturned her confidence in her plans and called her ideas of how life would evolve into question. Smoke filled her lungs, leaving the realisation that she couldn't depend on the air being clean anymore. Performing on stage in Sydney, she was preoccupied by overwhelming thoughts that she'd be caught in the city as her tiny family was being evacuated from their home 60-odd kilometres away. There was the constant terror at night too.

'That's probably where this huge anxiety started for me,' Stone said. 'I think I've probably experienced anxiety all my life, but I didn't have a name for it. It just whipped up; it took a very strong form of an existential crisis. I was just completely freaked out.'

A return to the United States to continue the pursuit of her acting career was imminent, but Stone couldn't shake the responsibility she felt from the hefty carbon load she carried from frequent flying. Frantically searching for low-carbon bandaid solutions that would support the long-awaited plan to split her life between Australia and Austin, Texas, Stone evaluated some out-there options, but cruise ships and yachting across the ocean Greta Thunberg–style were quickly discarded. 'I suddenly realised that this is a life-reconfiguring question; it's bigger than finding just a quick solution.' A couple of weeks later, after confronting some pretty intense nerves, Stone had the conversation with partner Jack that would change the course of their life together forever, and begin to calm the emotional whirlpool Stone was caught in. 'I remember we sat in the driveway of our house and I just said, "I don't think I can keep living like this, it doesn't feel right." And because he is the person that he is, he was completely open to that.'

Saying those words began a transformation that Stone is discovering, but no matter where things lead, there's no question she is 100 per cent all in when it comes to climate action. 'That was the start of the journey and I'm still really, really early on in this whole thing, but I have to take it really seriously,' she said. 'We don't have the time, so let's just go great guns, let's just try to do this.' Stone decided as a first step to connect with her community and quickly learned how important coal is to the history and culture of this place, beer and black diamonds aside. 'This is people's history, and history is important and identity really important,' Stone said. 'What do coalmining and steelmaking mean? They mean unionised jobs, protected jobs that have good benefits, that put food on the table for families and that are inherited over generations, and there are all the industries that spring from coal-mining activity too.' She added:

> I started to see that I've been really judgemental. I'd blown in from Sydney, via seven years in New York, and it's not my place to come in and judge the history. It's my place to try to understand it, to integrate, to learn about the stories and work out what is this complex relationship between the future and the past.

Stone started making a documentary film, which has evolved into a creative approach to connect climate action to practical support for local transition. The friendly name of the project, Hi Neighbour, embodies the approach, which aims to protect workers and remove obstacles of transition by building community energy projects and directing profits to a fund that supports retraining opportunities for any local worker looking to move into renewables. 'Is there support for those workers? Not really,' Stone said. 'We give up our paper coffee cups and they give up their jobs.' When I caught up with Stone, her idea is gaining traction in the Illawarra, with big industry, unions, councillors and workers interested in talking and the organisation newly registered as a not-for-profit organisation. 'I'm able to connect people with other people, and start to feel those branches grow, and that feels really empowering,' Stone said. 'Hopefully we can break down that rhetoric and really damaging politicking that's been built up over years and start to have a vision of a future that's safe and protected for everybody.'

Hi Neighbour emerged from Stone's connection with a place that she intends to be her forever-home. 'When you invest more in a community, you get back more from the community,' Stone said. 'That's a great gift of being in one place.' Other rewards are many: Stone reflected on how she had become closer to her extended family after being away for more than a decade, how her hands were closer to the soil, seeing her garden grow, the seasons shift, and knowing when the winds are changing. One pathway has closed, but step by step she's discovering new tracks and building new skills along the way. 'The biggest thing is that rather than a load on my back, this has been a great way to free myself and see myself in a different way, rather than feeling limited by my understanding of what I'm capable of,' Stone said. 'What this waking up has given me is a realisation that I can make things, I can create things myself, and I can inspire people to join me. Rather than just like a victim of whatever's coming down the line, I feel part of trying to change that tide.'

BOOM TOWN

I landed on Awabakal Country at five years old, brought to the banks of Lake Macquarie when Dad was transferred from his job with the State Electricity Commission at Lithgow's Wallerawang power station to Eraring power station, one of Australia's biggest carbon polluters. Not that I knew it at the time, of course; I was too busy learning how to fall in love with the salt water, my skin scorching beetroot red in minutes as I jumped off boats to collide with the lake's massive jellyfish that made us yell with fright as they brushed against our legs, propelling us to swim faster. I remember at around fifteen years old, a bunch of us took to jumping from a mean height into the nearby Munmorah power station's outlet canal, the warm water gently washing us back into the cool shallows at Lake Munmorah's edge. That was back when a job for life was heavy industry and manufacturing: 12,000 steelworkers at Newcastle's BHP steelworks dominated the city centre; 1000 people worked at the peak of the Hunter textiles operation at Rutherford, and a clutch of mines and power stations made everything work. The city was bustling, until one day it wasn't, like so many places that live through the boom-and-bust cycles of heavy industry.

BOOM TOWN

Our town, is a boom town
but you'd never know it was a boom town 'cause
the main street is empty and the shops have all closed down
There are squatters in The Empire.
That old seedy late night bunker
and though I barely went there
I still miss that pub like all the ghosts
of our town.

—'Boom Town', Steel City Su (Su Morley)

The steelworks and textiles are long gone, but in the two decades since BHP's closure sent shockwaves through the community, Newcastle has risen to claim the unenviable title of the world's largest coal export port. Around 160 million tonnes of coal was shunted through the mouth of the Hunter River in 2020, more than what was used by all of Australia's power stations combined. It's where climate activists have campaigned over decades to hold back the coal trains, prevent mine expansions and stop coal-seam gas in its tracks, where workers have fought to hang on to their prospects for good, local jobs, and where politicians of all stripes have worked to keep the climate and energy culture wars alive and kicking for short-term power-plays of their own, divided communities the collateral damage.

Steve Murphy knows this story too well, from his early days in the Hunter to his role as National Secretary of the Australian Manufacturing Workers' Union. 'This is the part of New South Wales that is the source of all wealth, but if you drive through it, you wouldn't think that billions and billions of dollars are dragged out of the ground every day,' Murphy said. 'Very little goes back to the community.' Growing up in Telarah, Murphy was lulled to sleep listening to the coal trains rattle through the Hunter Valley to the port, the soundtrack to a place where people work to keep the state's industry and national economy thriving. 'A coal worker doesn't wake up in the morning, looking forward to polluting the environment,' Murphy said. 'Workers wake up in the morning to go underground into this dangerous environment to keep everybody's electricity working and to feed the family. So when a worker's driving to work in the dark and driving home in the dark and looks up and sees the streetlights on, they say, "Yep, I did a shift of sacrifice but I kept those on for every family in New South Wales."'

Watching the culture war raging in the Hunter, Murphy had the realisation that something had to give after the 2019 so-called 'climate election'. The front pages of newspapers spouted commentary that blamed the Labor and the Greens parties' climate positioning as a major driver of the result, and then recommended they needed to abandon its centre-left values to appeal to working people. Murphy called bullshit, but was asked by fellow unionists as much as questioned himself: how are we going to have this debate, to neutralise this war that is going on? 'We saw workers in these blue-collar regional areas throwing rocks at environmentalists,' Murphy said. 'I guess the reflection was you can have the best policies in the world, but if you don't bring people with you then it doesn't matter a crack, right?' Rather than lob his own opinions into the post-election media ruckus, Murphy decided to do some deep listening, first by bringing people together for wide-ranging discussions that unpacked the many challenges working people faced across the nation, especially in the Hunter. Labor Environment Action Network (LEAN)'s national co-convenor Felicity Wade was invited, and she took time to reflect with the group on the long history of environmentalists and unionists working together for justice, a fair share, for place and people. 'When working-class politics gets ahead of the environmental movement we win,' she told the meeting. (Wade and Murphy dispute who said it first, but Murphy insists it was Wade's wisdom that cut through.)

One of the union's delegates stood up, telling his comrades that, although he was in the industry, the job paying the mortgage, he wasn't particularly loyal to coal. 'He said, "if you build me a job that I can go to tomorrow that pays around the same money, then I will shift, I'll go tomorrow",' Murphy said. 'That gave me this idea that this was absolutely doable.' Murphy then spent weeks canvassing through the Hunter, looking up mates from his early days as a fitter and turner on the railways and at BHP's Newcastle steelworks. Old friends and new told Murphy three things: first, that they accepted change is on the way; second, local workers did not trust politicians or mining magnates to look after their prospects, and; third, unions are trusted but weren't doing nearly enough. 'So I had a cup of coffee with Felicity and said, "I reckon that there's something in this; we've got to plan for the future, and I reckon the best way for us to do it is to work together,"'

Murphy said. 'We started to draft up the kind of things that we would all agree on and we agreed on the fundamentals: the union will still go and do crazy stuff and LEAN will still go and do their crazy stuff. But in the middle, there's a lot we can cooperate on.' An idea of collaboration began to take shape.

LEAN's Felicity Wade spent fifteen years with the Wilderness Society, winning big combative-style campaigns that were all attention and pressure in the name of nature protection: she dropped banners off Sydney's Harbour Bridge and was arrested occupying former Prime Minister John Howard's office in 1996. Fights were big, but wins were bigger: Wade's work was instrumental in securing protection for hundreds of thousands of hectares of forests and wilderness in New South Wales, requiring bloody battles with the logging industry, the agricultural industry and 4WD lobby. Death threats were part of doing business. 'Our slogan was "wilderness, no compromise",' Wade said. But climate change demands consensus building across all sectors of society, Wade now argues, primarily because of the sheer scale of the climate challenge in front of us, but also to ensure climate doesn't fuel toxic politics of the kind that created Trump and Brexit. Wade sees One Nation's chunky 21.8 per cent slice of the primary vote in the seat of Hunter at the 2019 federal poll as a warning bell that warrants our attention. 'The right's effective weaponisation of these fears has been a major brake on climate action in Australia and the problem threatens to escalate,' Wade said. 'Often what these communities hear from those of us who care deeply about this issue is, "your job is not important when the planet is in peril". This attitude is the root of much of our failure to build a consensus in Australia on the need to act.'

While climate change can be addressed through shifts in the deployment of private capital and visions of billionaires, Wade argued that without government intervention to make these changes fair, social dislocation would be immense and our cohesive society be put at risk. 'We have to be part of conversations in mainstream institutions; we have to build bridges and understanding,' said Wade, who made a clear-eyed decision to work 'in between', leading environmental advocacy within the Labor party until, after the 2019 election, she decided to focus on communities like the Hunter. 'LEAN being trusted by the union movement allowed me to make the connections and create something that holds more energy, more local legitimacy and greater power for change than either of its constituent movements acting alone.'

COME TOGETHER

George Woods is a lifelong greenie who grew up in Newcastle, fascinated by the diversity of the catchment she described as a transition zone between the Sydney sandstone in the south and subtropical rainforests of the north-east of the state, the Barrington plateau's World Heritage Antarctic beech forests in one corner, the Wollemi wilderness, also World Heritage listed, in another. 'It all just pours down here to where I am, and I'm right next to the estuary, which is this internationally significant wetland with birds who fly here from China and Korea and Japan every year,' Woods said. 'It's everything about this place, I just feel attached to it.'

Woods has been part of Lock the Gate, a grassroots alliance that's committed to delivering sustainable solutions to our nation's food and energy needs, for close to a decade, fighting coalmining expansions, and fending off coal-seam gas. Five years before Steve Murphy had those first promising conversations with Hunter workers, Woods was searching for a new way in the region after years observing government-led approaches that simply weren't working. 'We work with farmers and people in quite small rural villages who tend to be the ones who get affected by the mining projects,' Woods said. 'In town tends to be where people are in favour of mining, because they're small business people, real estate agents, the publicans.' Public hearings on coalmines in Singleton or Denman were more a theatre of contention and irreconcilable difference than debate or community consultation: the farmers and horse breeders sat on one side of the local hall, the miners in high-vis gear on the other. The two groups came together only when pleading for their respective industries. So Lock the Gate decided to do something different, activists knocking on doors in the bigger towns of Singleton and Muswellbrook, searching for any potential place to connect. Surveying locals, Lock the Gate members were not surprised that agreement flagged at around 40 per cent when they were asked whether the industry had done more harm than good or whether there should be no more coalmines approved. But then the breakthrough emerged: nine out of ten people agreed that the region needed a plan for the future. 'We saw that as a chance to get away from this really quite unproductive, obstructionist dynamic where we were having to fight against people saying that they were fighting for their livelihoods, and actually create a different dynamic in the region,' Woods said. 'We wanted to find a place of common ground where we could talk to a broader number of people in the community.'

This was the beginnings of Hunter Renewal, a unique project that started with dinners of 100 people where locals were served Hunter Valley produce and gently guided to discuss what they valued about the region, what they loved, what they held in common. 'In my experience, all people, no matter what their relationship is with the mining industry, really love the beauty and the lifestyle, you know, like they love living in the Hunter, they love the community,' Woods said. The taboo in discussing where the community was headed seemed caught up in a collective fear of tempting fate: talk about the end of coalmining and you might trigger the collapse of the industry. Rather than push back, Lock the Gate made the commitment to people and place, dedicated time to listen and allowed conversations to evolve. Before too long the space began to grow, with local councillors and businesses getting involved in the project. 'I think it was a pleasant surprise for many that it was not hard at all,' Woods reflected, her smile getting wider. 'People say to me, "it must be really challenging hammering out agreements", but it's actually not very hard at all because we really are coming from the same values. We've done a lot of outreach over the five years, and we felt confident, increasingly confident, that there was broad agreement across the region among almost everybody that there needed to be a plan for diversification and adjustment.'

Fast forward to 2020 and Lock the Gate is a member of the Hunter Jobs Alliance (HJA), a collaboration that emerged from Murphy and Wade's conversations, where thirteen unions and environmental organisations stand together to work with governments and industry to deliver a safe, prosperous future for the Hunter, one in which workers, their families and the environment can continue to thrive. 'The union movement and the environment movement coming together through something like HJA is pretty unusual, but it should be far more common, of course,' said Chris Gambian, chief executive of the Nature Conservation Council of New South Wales, an HJA member. 'We want to partner with folks who care about making a real contribution to climate action, and this seemed like something where a lot of good faith had been built up after a lot of effort.' When I asked why greenies and workers should get together,

Gambian, a former union organiser himself, reminded me with a chuckle that greenies and workers are often the same people, and climate change is too big a problem for any of us to solve alone. 'I don't think it is possible to genuinely address the threats to the environment without a deep understanding of how that destruction interacts with the economic model. By working with others who have a stake in the economy, and creating a shared vision for the future in which nature and people can thrive, I think we make ourselves so much more effective.' Woods saw the HJA is fundamentally an exchange between people, with open-mindedness and a generosity of spirit as the characterising value. 'There's this sort of righteousness that tries to carve out for itself a moral high ground from choosing not to see the predicament that other people are in,' Woods said. 'If a politician refused a coalmine and then they lost an election because of that, well, that's a genuine predicament that person is in, or a banker taking some risk with their money or whatever responsibility they have. They are in genuine predicaments. I think it's equally important for all of us to be open to understanding the predicament that other people are in and trying to help them get out of them.'

'People in the Hunter have a highly developed bullshit detector,' Wade says. 'Being a voice of calm and offsetting the extremes of the debate on both sides, speaking of the inevitability of change, talking of the urgent need for the region to work together to diversify, avoiding simplistic solutions to a complex problem—these principles drive the alliance.'

The three goals set by the brand-new HJA in 2020 were achieved in less than twelve months: establishing a legitimate voice that has directly and successfully challenged the region's climate and energy culture wars; securing a $25 million fund from the New South Wales Liberal–National state government to help workers and communities shift to new livelihoods; and getting Tomago Aluminium, which is around 10 per cent of New South Wales' electricity demand, to commit to switch to renewables— the company is on board to do that by the end of the decade. 'It was a surprise that once we set our minds to it, how quickly we could have those discussions within the political sphere, not just with Labor or the independents, but even some of the people within the National party were secretly saying this is a worthwhile discussion,' Murphy said. Next on the agenda is to ensure benefits are shared across the community, including world-class education, fully funded community services that can help any worker who is in transition or families who are struggling to find work, and securing jobs for young people across industries in the Hunter. HJA is also pushing for a fully funded, regionally led Transition Authority and carefully calibrated industry policy to ensure the Hunter remains a manufacturing powerhouse with opportunities for all.

The Just Transition Declaration launched at the United Nations Climate Conference in Scotland in November 2021 is a statement of support for delivering sustainable, green and inclusive growth to decarbonise our economies, in line with limiting global average temperature increase to 1.5 degrees Celsius. The declaration highlights the importance of support for workers in the transition to new jobs, support for social dialogue and stakeholder engagement and local, inclusive and decent work. Sixteen nations, plus the European Union, supported the declaration but Australia didn't sign on. Communities are refusing to sit back and wait for the policy vacuum to be filled, instead coming together to develop DIY industry policy through thoughtful listening, building networks and generating ideas far from the boardrooms and halls of power. 'We're largely kind of getting that consciousness there about this idea that blue-collar workers can start to determine what their future looks like, and we just get the sense from the environmental movement that finally they're starting to see the light at the end of the tunnel,' Murphy said. 'Not only are we going to take action on climate change, but we're going to have job creation and we're going to have justice for working people as part of it as well.' Efforts to build connections and consensus are emerging in the Latrobe Valley in Victoria, in Collie south of Perth in Western Australia, and in the epicentre of noxious climate politics, central Queensland. Who knows, perhaps we will see a national jobs alliance emerge from the ashes of our climate debate, uniting communities and demanding leadership from our political class.

POSSIBILITY AND PURPOSE

Internationally renowned environmentalist and entrepreneur Paul Hawken wrote that the ongoing cause of degeneration is inattention, apathy, greed and ignorance, which reads like a playbook from the last two decades of Australia's climate wars. But Steve Murphy and George Woods speak of each other in such glowing terms. They show us what's possible when a challenge is shared. 'George Woods is amazing, she's an absolute rock star,' Murphy said. 'We had to walk through a lot of uncomfortable conversations, but I just have the utmost level of respect for George and what she does. I probably wouldn't endorse most of the tactics, but in terms of the thinking and the vision, and just the energy that she brings, she's fantastic.' Woods, a self-described 'divisive figure' in the Hunter over the years, said she has a lot of love for Murphy too, for his catalytic role in the HJA's establishment and taking the risk of including environmentalists, and for the integrity and sincerity he brought to the task. 'Something you go to and fro about in advocacy work, is how much is it about particular people and their talent and their character, and how much is it about an organisation or a group effort,' Woods said. 'I'm a big believer in group effort, but sometimes there are these extraordinary people and Steve is one of them.'

When I asked Murphy about what he'd learned about consensus, he framed the lesson around what's most important to all in the HJA. 'Fundamentally we all want justice, we all want decent wages, we all want good, just, secure jobs,' he said. 'We all want to be able to live in a house where you feel safe. We want to have a sense of community about where we live; we want there to be justice in the world. Sometimes we disagree on the detail, we might disagree on the tactics, but fundamentally we've got the same values. Our organisations function in all the different weird, clunky ways and you scratch your head and go, "How do you make that work?" Then you walk through a lot of uncomfortable conversations, and it's the energy that we both bring—it's that old adage that you're stronger together.' I was reminded of my stroll around Green Connect's farm, Flament critical of our society's tendency to put people, projects and organisations into silos, looking at fragments rather than the system as a whole. 'Every time we consider doing a new project, or which way to go within Green Connect, we ask ourselves, essentially, three questions. "Is it good for people? Is it good for the planet? And is it financially sustainable?" And if the answer is yes, we do it,' Flament said. Questions to live by.

TOGETHER WE CAN . . . *find common cause*

✳ Permaculture principles can be applied beyond the market garden: concepts of people care, Earth care and fair share are an ethical framework that we can use to guide how we can live our best lives at home, at work or in our communities.

✳ Industrial regions around Australia are forging new pathways of connection and demonstrating how building consensus can help to ensure a fast and fair shift in our economy.

✳ If we're going to get the job done, we need to remove the silos in our communities and look holistically at the challenge and potential solutions.

✳ Finding shared values and being willing to take a risk are key to seeing communities make the most of changes to our economy and society that are coming at us, fast.

FURTHER READING

✳ Green Connect Illawarra, Green Connect: Jobs for people and planet, 2022–, viewed 20 February 2022:
https://green-connect.com.au

✳ Hunter Jobs Alliance, 2022–, viewed 20 February 2022:
www.hunterjobsalliance.org.au

✳ Lock the Gate Alliance, Hunter Renewal, n.d., viewed 20 February 2022:
www.hunterrenewal.org.au

✳ Moloney, Hannah, *The Good Life: How to grow a better world*, Affirm Press, Melbourne, 2021

✳ Permaculture Principles, n.d., viewed 20 February 2022:
https://permacultureprinciples.com

✳ United Nations, 'Supporting the Conditions for a Just Transition
  Internationally', UN Climate Change Conference UK 2021,
  4 November 2021, viewed 20 February 2022:
  https://ukcop26.org/
  supporting-the-conditions-for-a-just-transition-internationally

Part Two

PROJECT NOT PANACEA

CHAPTER 6

REFRAMING OUR THINKING

'Another world is not only possible, she is on her way.
On a quiet day, I can hear her breathing'

—ARUNDHATI ROY

It's a steamy day on Kombumerri Country, a part of the Yugambeh language regions known as the Gold Coast, Queensland, just the type of weather that would have you reaching for a cold one. I'm in conversation with Clinton and Lozen Schultz, founders of Sobah, an Aboriginal-owned and led non-alcoholic beer company that's changing the way we see the humble brew while helping to build Australia's native foods industry. 'I want to see the custodianship and First Nations–owned businesses really excel in the native food space,' Lozen says. 'We've had a big impact in not just the non-alcoholic drinking space, but even in the agrifood business, in the native food space, it has also been quite significant.' The pair are frantic with pre-Christmas orders, the conclusion to a bumper year that has seen tripled production of their beers flavoured with finger lime, pepperberry, wattleseed and Davidson plum, and a successful ten-day $1 million crowdfunding campaign under their belts to allow the operation to keep up with ballooning demand. We're talking carbon, sustainability and social impact because I'm trying to understand: how does a small, but growing, business make a positive difference, environmentally and socially?

After Clinton Schultz's sobriety journey left him frustrated because he couldn't find a decent non-alcoholic beverage for himself, Sobah began its days in the Schultzes' garage, where they brewed beer to add to the menu at their food-truck side business that specialised in bush foods: think crocodile and kangaroo skewers and macadamia satay on emu. In just under four years, Sobah evolved into a fully-fledged social enterprise that aims to be carbon negative within five years, while pursuing a big expansion. It has been a journey full of challenges, including searching for better energy options and diving into the murky intricacies of water science, but the overall direction of travel is clear. Schultz, a Gamilaroi man and practising psychologist, said Sobah has been grounded in the philosophies of Gamilaraay Lore *dhiriya Gamil*, which include respecting people, place and the environment, a perspective of reciprocity in the way you deal with others. Ensuring positive change through promoting healthy lifestyle choices, social equity and sustainability has been prioritised, along with smashing stereotypes, uniting people and raising positive awareness of Aboriginal and Torres Strait Islander culture too. It sounded like a huge bundle of responsibilities to pack into a small can, but when it's boiled down, as Lozen Schultz pointed out, the overall approach was pretty simple: *Do less harm and do more good.* This is the core values set Sobah lives by, helping to guide daily decision-making, big and small, across the thriving operation. For the Schultzes, while beer exists to help make a living for them and their team, the product is an important vessel for building a deeper understanding of the strengths, history and practices of First Nations culture, through education and experience. 'It's really about trying to help people to understand connectedness beyond just their partner or their kids, and see it as a much more holistic concept,' Clinton Schultz said. 'That allows for healing for everybody in the country.' He added:

> I really work from all the foundations of my way of knowing, in all the businesses that we've run and in everything that we promote, because I think that they're good for everybody, not just us First Nations. We've survived on this country and managed this country really well for 100,000-plus years. We know how to do this; we've done it before. It's all in our story; it's all in our role. We have to return to Indigenous ways of knowing and being if we're to remain on this planet.

At Sobah this holistic approach was built into the everyday, where there have been conscious choices made that put what is best for people in collective and place, across time, ahead of what's best for 'me', the individual, right now. It's a small pause, Clinton Schultz explained, a thought process that usually takes only a few seconds to consider the choice you will make. Simple, yes, but it feels like a small pause is a big challenge to our industrialised consumer world of the last hundred years or so, where instant gratification reigns supreme. 'When you take the time to step back and reflect, you will come up with a better option,' he said. The Schultzes didn't use the words, but what shines through their orientation to their mission in work and life is regeneration, weaving together healthier food, education, social justice and community, recognising interdependence of the physical, social and cultural aspects of life.

Regeneration might sound like just another buzzword, but it's fast becoming much more than a shiny new label to replace sustainability; it's an approach that is increasingly being taken up by business leaders, industries and communities around the world. Paul Hawken defined regeneration as 'putting life at the centre of every action and decision'. 'If putting the future of life at the heart of everything we do is not central to our purpose and destiny, why are we here?' he asked. 'Regeneration is not only about bringing the world back to life; it is about bringing each of us back to life. It has meaning and scope; it expresses faith and kindness; it involves imagination and creativity. It is inclusive, engaging and generous. And everyone can do it.'

Regenerative practices see the do-less-harm approach of conventional sustainability approaches as not nearly enough when we consider the monumental tasks we have to build more capacity, resilience and healing to improve the quality of our lives in all dimensions. Regenerative change, wrote expert in the emerging field Carol Sanford, is built through taking conscious charge of our thinking processes and helping other people to do the same. 'Without the exercise of conscious awareness and choice, we are condemned to endlessly cycling through old, predetermined patterns, making it nearly impossible to transform ourselves or our world.'

Sanford, an executive educator of Mohawk ancestry who has spent more than four decades evolving her regenerative business design approach with Fortune 500 companies and new economy executives, wrote that regeneration has its basis in the science of living systems. 'We humans are all works in progress, we can continue to grow and evolve throughout our lives,' Sanford said. 'But to do so requires a certain kind of mental discipline. We must challenge the many fixed beliefs we hold about who and how we are.' To work regeneratively, Sanford argued, we need to consider wholes, a person, an organisation, a community or a planet, rather than break entities down into individual component parts. We also need to work from the basis that every entity is unique and has potential that can be realised, rather than spending all our energy focusing on the endless list of problems. I think about my work to build social justice over two decades, from advocating for freedom of the press to standing up for disability rights, campaigns to defend clean energy policy and more: it's work I am very proud to have the privilege of being part of, but it was certainly broken down into component parts; endless lists of problems were—and are—defined before looking first at wholes and solutions. What's more, I used to separate this work from the rest of my life: like keeping it in a locked 'work' box possibly so I'd have regular refuge from the intensity of it all. But joining the increasingly diverse community of climate advocates has, over time, shifted my thinking profoundly. The more I've embraced this community of shared passion and mission, the more comfortable I am becoming with seeing the roles I play in different parts of life: parent, manager, friend, sister, colleague, confidant, student, child, mentor, ally—and how they are more integrated than I thought. Of course there are moments when I need breaks as the ebb and flow of life overwhelms one or all of these roles at times, but, as I reflect on this, the more I've integrated the work on climate change as part of my life, I'm discovering life is becoming richer, and what I really value in life is becoming clearer too: deeper relationships with family and friends, being part of a more connected local community. Back at the office, I'm finding relationships with my colleagues are becoming closer and far more honest: work feels more like a place where I choose to direct the benefits of my privilege and my stores of energy, rather than it being some kind of 'feel-good' gig with a pay cheque.

COMMON GROUND

Investigating regeneration quickly leads me to the farm gate, where new ways of considering how food can be produced to benefit nature are under intensive exploration. It's no surprise why: Australian farmers are on the front lines of climate impacts, with increased frequency and severity of drought, floods and fires creating increasingly unpredictable conditions for cropping and grazing operations. Meanwhile, consumers like you and me are demanding more climate-responsible products and companies are responding, beginning to seek out solutions for their supply chains. This emerging and complex context is on top of the need for the agricultural sector, responsible for around 14 per cent of Australia's greenhouse gas pollution, to urgently reduce emissions. That's why I looked up Alasdair MacLeod, a media executive turned grazier who is spearheading efforts to boost 'natural capital' through our nation's and the world's farms. MacLeod, son-in-law of media mogul Rupert Murdoch himself, had a twenty-year career at NewsCorporation in Sydney until 2010: some of those years I was working at Australia's media union, representing journalists at the empire's HQ in Sydney's Surry Hills. Sitting at my makeshift office in the lounge room at home speaking to MacLeod who is Zooming in from his part-time base in Oxfordshire in the United Kingdom, I wonder to myself: could our divergent career histories find a convergence point in the present: is there anything we could possibly agree on? I'm reminded of the consensus-building that's emerging in the Hunter when we find common ground in the first few minutes of our conversation, MacLeod's passion for proving that healing the land, reducing emissions and producing food to feed all of us shining through. A few weeks after we spoke, MacLeod took his mission for improved soil health to the United Nations Climate Change Conference in Glasgow in 2021, reinforcing findings in the most recent IPCC report case that 80 billion tonnes of carbon, or 300 gigatonnes CO_2 equivalent, could be returned to the soil through improved farm management practices.

The shift from media to agriculture began when MacLeod and his young family spent school holidays at the family's wool-growing property near Yass, in the Southern Tablelands of New South Wales, where he observed firsthand the multiple crises that were brought about through the farm mismanagement during the long years of the Millennium Drought. Looking around the farm, he saw a decimated landscape with barely a blade of grass. 'It was what I like to call a sort of perfect storm of crises,' MacLeod says. 'The farm's manager created an economic crisis because we had to spend a heap of money buying in feed for the animals.' Added to this were the animal and human welfare crises as well, MacLeod said, with stock dying and soaring stress levels affecting everyone working there. 'What happens when the farmer doesn't anticipate drought and doesn't manage it, it's disaster.' Witnessing this blow-by-blow devastation, MacLeod started searching for answers around the same time interest in regenerative agriculture began gathering pace through farming communities in regional New South Wales. 'We weren't really talking about carbon in those early days, it was about climate resilience,' MacLeod said. 'It was about making sure that you were managing things in such a way that, when dry times hit, your system was resilient enough to manage its way through that without creating the destruction that I'd seen on our property.'

Visiting field days, connecting with regenerative leaders Charles Massy and Terry McCosker, were influential for MacLeod; so was connecting with Soils for Life, an organisation founded in 2013 by Australia's former governor-general Major General the Honourable Michael Jeffery, Australia's first National Soils Advocate. The organisation showcases farmers who demonstrate deep under-standing of how to manage landscapes for climate resilience while keeping properties productive and profitable. 'It was pretty exciting to think about how I could contribute to a new way of approaching land management and a new way of looking at how agriculture could manage its way through increasingly difficult periods with great climate variations,' MacLeod said. 'It was only later that I began to realise that, if done properly, agriculture could be part of the solution rather than part of the problem.'

MacLeod started out by building up the Wilmot Cattle Company, a cattle trading operation spanning 5665 hectares on Kamilaroi, Gumbaynggirr and Nganyaywana Country in the New England region of New South Wales, which makes it a mid-sized farming operation by broad Australian standards. 'Our goal is to still produce as much beef as we have in the past, but with less inputs, and in a more ecologically sustainable fashion,' explained Wilmot's general manager, Stuart Austin. A whole-of-ecosystem approach is taken across Wilmot's three properties: Wilmot, Woodburn and Morocco. Soil nutrients are tested annually, and assessments made of solar and water resources to safeguard the system effectively. How does it work without all those phosphorous-based fertilisers? 'We're effectively trying to mimic how Mother Nature used to manage the landscape,' Austin said. 'Imagine the large herds of buffalo running across the plains in North America or the herds of antelope in the savannas in Africa, which were very migratory: they were moving across the landscape in very tight mobs to protect themselves from predators, briefly grazing an area before moving on, following the seasons and not coming back until country was ready to be grazed again.' Wilmot uses fences to mimic these natural patterns, shifting cattle around small pastures frequently, sometimes as often as twice daily, so the pastures can recover. Close to a decade has been spent at Wilmot improving pasture, with management practices seeing ground covers and biomass build up and around 25,000 trees planted, providing shade for livestock and wildlife corridors for native wildlife. Pure organic playground it is not quite, with blackberries sprayed with pesticides every few years and a few other invasive species managed manually, but fertilisers that fuel most of our nation's industrial-scale meat production aren't needed here and almost anything that grows on the farms is free to flourish. 'There have been quite minimal inputs here, and what that's done is dramatically reduce overhead costs, which has seriously lowered our risk so we don't have to extract very high production at the farm to make a profit.'

Efforts to improve soil health helped Wilmot secure a $500,000 carbon credits deal with Microsoft in 2021, a deal that aims to demonstrate the potential for farmers in Australia and around the world to sequester soil carbon for a return. It's a controversial area with evolving science of how to measure additional carbon captured in the soil through these and other methods and how to monitor and set standards for maintaining properties that receive credits. There is also, of course, the challenge that carbon may be difficult to maintain in soils over time. But as regenerative agriculture proponent Terry McCosker reminded me, we don't pay the true cost of the food we eat right now: the environment pays for us, and pays dearly. 'If we're going to change that, there has to be a way that the rest of the environment is paid for as a public good,' he said. 'I believe that soil carbon credits and biodiversity credits coming back into the landscape is very important and will get a job done that no other sector can do.' McCosker and MacLeod have a point here: if we want to look holistically at systems that have potential to help combat climate change while restoring ecosystems, we should give it a go.

McCosker, who delivers training and support services through his company Resources Consulting Services or RCS, has spent the past 30 years evolving the specifics of regenerative agriculture practices, with a clarity of vision that hasn't changed much in that time. 'We support farmers, the world's most honourable profession, to produce quality food which produces healthy people. That's the simple version of it.' Interest in the field is accelerating, McCosker said, with thousands of farmers across Australia now trained in regenerative practices, many looking for ways to become more resilient in the face of a more uncertain climate. Meanwhile, big players in beef supply chains have been in touch, looking for ways to tap into regenerative agriculture markets and, in 2021, two of Australia's largest banks, Commonwealth and NAB, piloted new green loan products that provide lower interest rates for producers engaging in more sustainable farming that improves biodiversity and soil health, protects waterways and sequesters carbon. 'Markets and consumers want to link back to farmers that are doing the right thing for the environment; we see those things coming together very, very quickly,' McCosker said.

MacLeod's family foundation, Macdoch Foundation, has kicked off a new program called Farming for the Future, described in its first discussion paper by PWC as aiming to identify pathways and develop a decision-making framework that will deliver long-term economic, environmental and social benefits from farming that improves natural capital on-farm, at scale. The paper notes that a recent Australian study over a ten-year period found that graziers investing in natural capital were more profitable than similar farms in their local area. Farms that enhanced their natural capital, the report noted, could see average annual net income increase between 40 per cent and 83 per cent based on studies conducted in Australia and the United States. It's early research with small sample sizes, but the PWC report put the case that there is enormous potential for farmers to improve the soil, water, remnant native vegetation, environmental plantings and animals to deliver climate-friendly outcomes from the nation's agriculture sector while improving resilience of farms and farmers to crisis situations. 'This is not just a hippie farmer movement, this is not just for people with little hobby blocks,' MacLeod said. 'This actually is a way we can create a resilient, profitable productive agricultural sector, while at the same time deploying that sector to solve the emissions issue.' When we spoke, MacLeod noted that the National Farmers Federation had just signed up as a program partner, in a sign that system conditions could be shifting:

> I think it has been a process of moving from one mindset to another mindset, and we're really comfortable with where we are at the moment. Our businesses are more profitable than they've ever been; our businesses are more resilient than they've ever been. The people that work on the properties are more, you know, fulfilled and satisfied with what they're doing, and we're beginning to understand the benefits of doing this in a way that you can monetise.

Josie Warden, Head of Regenerative Design at London's prestigious RSA (that's the Royal Society for Arts, Manufactures and Commerce in the UK), wrote that a 'regenerative' mindset is where one sees a world where living beings and ecosystems rely on one another for health, where reciprocity and co-evolutionary relationships abound and we are shaped by and are shaping our connections with one another. 'Interdependencies and relationships have never been more important,' she emphasised. 'The social challenges that we are grappling with are nested within our environmental ones, carbon emissions are intertwined with community health, biodiversity with social justice, and so on. The world is made up of living systems that are complex and emergent, not linear and predictable. But humans are hardwired to thrive in this world and the potential to act is already within us and our communities.' As the principles and actions in Table 6.1 show, there are a number of ways we can start to act on our own potential to bring about regeneration as soon as we decide to.

Regenerative thinking, Warden wrote, recognises the interconnectedness of the social and environmental challenges we face, with an understanding that we need to rebalance and restore these relationships. 'Designing regeneratively involves a developmental outlook and requires us all to work on ourselves and our mindsets and behaviours as much as on the infrastructure, institutions, services and products in our external world.'

FRAME CHANGE

This way of seeing the world isn't anything new; it's more something we've lost as technology and economies developed and our individual, competition-based economy has come to dominate our culture. It doesn't take much effort to see more interconnected

TABLE 6.1 ACHIEVING REGENERATION

Principle	Launching point	Diving in
Start with place and context	Recognise that people, places and communities have different and unique qualities.	Question assumptions that context-agnostic or top-down solutions will work in any and every place.
Seek different perspectives	Regenerative thinking recognises that complex problems look different from various perspectives and that a diversity of views are needed to address them.	Ask what the blind spots might be and how they might be illuminated.
Build capability and reciprocity	Work with people and places to create shared ownership of challenges and find shared solutions.	Support others to build capabilities and nurture relationships, mutuality and reciprocity, going beyond transactional 'you scratch my back and I'll scratch yours' exchanges to more systemic interactions.
Take a nested systems view of success and consequence	Look beyond financial value and narrow measures of success. Recognise that you are working with nested wholes and be aware of the relationships between layers.	Always think about the impacts, consequences and contribution of your work on the wider wholes, both intended and unintended.
Design for circularity and circulation	Ensure information, value and power, as well as physical resources and elements, can flow across and between layers of the system so the entity regenerates.	Enable participation and actively engage and create spaces for the exchange of ideas; encourage plurality and diversity.
Create space for emergence	Test and iterate ideas and activities, rather than planning then acting at scale.	Ask how you might cultivate an experimental culture, creating space for questioning assumptions.
Design from a hopeful vision of the future	Working from a place of hope, the 'what if?', can build energy, momentum and commitment for the work that needs to happen now to realise it.	The future is not predetermined, so use hopeful visioning to move beyond short-term barriers.
Work on the inside as well as the outside	Remember that your interior conditions—how you think, reflect, communicate—affect everything you do.	Consider how your perspectives are changing, and how you're reflecting on these changes.

Source: Adapted from RSA, Principles for Regenerative Futures programme, 2021

communities and, frankly, more sophisticated ways of thinking, collaborating, working and potential-realising springing up, from the brewer's warehouse to the farmer's pasture. So why is it that we can sometimes feel stuck, considering the world's problems to be too hard to change? Cognitive linguist George Lakoff said it comes down to frames, or 'the mental structures that shape the way we see the world'. Frames are activated all the time from what we're exposed to, and a single word can be all it takes to activate and reinforce the way our brains have organised our worldview. If you encounter ideas or words that sit outside this worldview, they can be perplexing for your brain to comprehend. (If you've ever locked horns with an outright climate-denier, frames might offer some explanation on why you can't get through on facts or logic alone.) It can be difficult for people to get their heads around climate change, Lakoff said, because it is a systemic problem that requires systemic solutions beyond the frames already established in our brains: unfortunately, most of us haven't achieved a PhD in systems thinking by the time we've graduated from high school. So when we talk about 'reframing', we're working out how to shift the boundaries of what is considered plausible and possible. 'When we successfully reframe public discourse, we change the way the public sees the world. We change what counts as common sense,' Lakoff said. Our words shape the way we think and how our emotions play out, so the way we shape our language makes a difference in terms of what we think can be done and what will compel us to act.

Is it possible to change mindset fundamentally given entrenched frames are activated in our brains every time we hear a word or two? One telling example shows how language and people can shift is from United States communications strategist Frank Luntz. In 2003 Luntz advised the Bush administration to shift the term 'global warming' to 'climate change' because the latter is more controllable and less emotional a challenge: and it's harder for people to pinpoint who is responsible. It's hard to believe people make a career out of doing this degree of unconscionable damage, but what's interesting here is how dramatically Luntz shifted his own mindset on climate, and his recommendations on how we should talk about it. Almost two decades after penning that memo and, following wildfires that came achingly close to his Los Angeles home in 2017, he was sitting in front of a 2019 United States Senate hearing, saying he was wrong, climate change was real and action was necessary and urgent. Luntz then made a series of recommendations for climate change communications, stressing the need to jettison extreme language in favour of a practical 'get it done' approach. People want to know the positive, not just the negative, Luntz said. The recommendations align with multiple climate change communications research projects conducted in Australia in recent years. Erik Curran on Resilience.org wrote of how Luntz stressed the importance of framing climate action as a 'no-regrets' strategy, a position reflected in Luntz's famous list of words to use and lose on climate he presented to the same Senate hearing:

Use	Lose
Cleaner, safer, healthier	Sustainable/sustainability
Solving climate change	Ending global warming
Principles and priorities	Values
Reliable technology/energy	Ground-breaking/State of the art
New careers	New jobs
Peace of mind	Security
Consequences	Threats/Problems
Working together	One world

Here's a two-minute experiment you can run right now. Look at 'cleaner, safer and healthier' above: these words activate the frames around good health and a protected environment as a virtue we desire, conjuring up practical measures like picking up litter or regular visits to the GP. The word to lose, 'sustainability', is confusing at best, or meaningless at worst for many; it can trigger more questions than answers. It's a similar story when it comes to 'solving climate change'; the phrase connects with the broadly supported idea that humans are good at figuring out complex problems, where 'ending global warming' seems entirely overwhelming to the everyday person. 'Working together' respects individual autonomy but highlights how it's useful to cooperate in an organised way, where 'one world' suggests everyone must conform to one way of being in society.

Above all, Luntz said, there is one word that is most powerful and compelling that we should be using far more often, and it's a good one: imagine. Imagine, Luntz said, is one of the most powerful words in the English language because it individualises and personalises communication. 'If I asked you the question "Imagine life at its best?", you don't hear my voice anymore, you hear yours,' he said. When I hear the word 'imagine', I think of the future I want, I feel excitement and it activates the belief I have in the creativity and problem-solving abilities of our species.

Making his case for building space for more imagination in all aspects of our lives, environmentalist and founder of the transition towns movement Robert Hopkins delved into the brain's hippocampus. Located deep in the temporal lobe of the brain, it's the place where we transform our short-term recall into long-term memories, where we process context, pattern separation and spatial awareness and, crucially, where our brain imagines. When ideal conditions are available for the hippocampus to function,

Hopkins wrote, imagination flourishes, but it is a fragile structure, vulnerable to the stress hormone cortisol that is a near-constant companion to modern life. If the hippocampus is damaged, we can experience everyday events as more stressful, we are more prone to view the future negatively and we may even seek out information that confirms a more pessimistic worldview. This in turn releases more cortisol, resulting in the vicious cycle taking another turn around the carousel, for the worse. 'The thought that keeps me awake at night is that the further we get into the big challenges of Now—economic inequality, climate change and the very real risk of the collapse of many of the key aspects of the economy we depend on, mass migration and so on—the less able we are to imagine a way out of them,' he wrote. But there is hope here too, because imagination is resilient. 'If we can put to one side the factors that are suppressing it, it will re-emerge, blinking into the light, because it is our natural state.' It sounds like a tantalising invitation for our minds to wander some new pathways and become more open to what's possible.

Environmental philosopher Glenn Albrecht created the evocative word solastalgia in 2003, which he defined as 'the pain or distress caused by the ongoing loss of solace and the sense of desolation connected to the present state of one's home or territory'. Putting it another way, he wrote: 'It is the existential and lived experience of negative environmental change, manifest as an attack on one's sense of place.' It's essentially a feeling of homesickness when you're still at home but the place has radically changed around you. Solastalgia has permeated our culture and society, with it being used across a range of academic disciplines and in the creative arts too: Australian musician Missy Higgins' album *Solastalgia* ruminates on the complex emotions that surround considerations of environmental collapse. Albrecht has, encouragingly, always argued that solastalgia is not necessarily a permanent state: it might only need a satisfactory restoration of place to bring solace and comfort to those experiencing the negative emotions.

A relentless, optimistic and practical view of the future sits at the heart of what Albrecht named the Symbiocene, a term drawn from the word 'symbiosis', which originates in the idea of the companionship of life, and '-cene', which means a period in the planet's history. The Symbiocene is designed, he wrote, to counter the 'ruthless pessimism from the critics of the Anthropocene', the definition of recent history, when humans began to have a significant impact on the earth's climate and ecosystems. Albrecht has also produced new words to describe an array of positive earth emotions, and for me soliphilia is the most compelling. Soliphilia is essentially a reversal of solastalgia, where love of place is combined with a responsibility for protection and conservation. 'The story of soliphilia is one of local and regional people responding to Earth desolation by political and policy action,' Albrecht said. These actions will result in places that are repaired and revitalised, which will lead to us experiencing more positive emotions that will sustain and heal us too.

Do you have an unconscious list of what's impossible or what's possible that you find yourself referring to when you're making decisions? Perhaps you have both. But imagine if the walls you've drawn around the potential for yourself, your family, your community, your country or world shifted or dissolved altogether: what could happen? Perhaps the way to begin for us climate freaked-out folks is as simple as making the shift from asking yourself 'What *can I* possibly do now?' to make the slightest change of mindset to 'What *I can* possibly do now.' It's clear to me that, as Albrecht told a Sydney TEDx audience in 2010, we're going to have to shift our mindsets on a massive scale to realise our collective human potential. 'If we're going to solve many of these globally significant environmental problems, if we're going to solve them at global, national or even local levels, we're going to have to sort out what's going on in our heads,' he said. After all, he continued, without a hint of sarcasm: 'The future could turn out to be very ugly for humanity—but then again, it could be brilliant.'

Reframing our thinking is going to require reframing the language we use, the frames we activate and, ultimately, the way we view the world. Regenerative change, Carol Sanford wrote, is built on the power of taking charge consciously of our processes of thinking, and helping other people do the same thing. Or perhaps, as George Lakoff said, the simple first step we can take is realising that the way *you* think is not the only way *to* think.

TOGETHER WE CAN . . . *see what's possible*

* We need to shift to viewing the world holistically, looking at systems and how they are interconnected, rather than breaking things down into component parts.

* Regenerative change is a process of taking conscious charge of our thinking processes and helping other people to do the same.

* Frames are how our brains organise the world around us, and single words can activate and reinforce these frames.

* Reframing involves recognising there are other ways of seeing the world and using language that connects on values that stretch across different frames so we can shift views of what can be done.

FURTHER READING

✳ Albrecht, Glenn A., *Earth Emotions: New words for a new world*, Cornell University Press, Ithaca, NY, 2019

✳ Curran, Erik, 'Messaging guru offers list of words to use and avoid to build support for climate solutions', Resilience, viewed 18 March 2022: www.resilience.org/stories/2019-08-05/messaging-guru-offers-list-of-words-to-use-and-avoid-to-build-support-for-climate-solutions

✳ Hawken, Paul, *Regeneration: Ending the climate crisis in one generation*, Penguin Books, New York, NY, 2021

✳ Sanford, Carol, *The Regenerative Life: Transform any organization, our society, and your destiny*, Nicholas Brealey Publishing, Boston, MA, 2020

✳ Sobah Non-Alcoholic Beverages, Sobah, 2022–, viewed 20 February 2022: https://sobah.com.au

✳ Soils for Life, 2022–, viewed 20 February 2022: https://soilsforlife.org.au

✳ Warden, Josie, 'What does "regenerative" thinking mean?', *RSA Journal*, 4 November 2021, viewed 20 February 2022: www.thersa.org/comment/2021/11/ what-does-regenerative-thinking-mean

✳ Wilmot Cattle Co, 2021–, viewed 20 February 2022: www.wilmotcattleco.com.au

CHAPTER 7

CREATING OUR FUTURE

'If a better world can happen, then we should make it.'

—KIM STANLEY ROBINSON

Meet *Hermetia illucens*, the black soldier fly, a trooper sent to attack global food waste. Its sleek body is far more elegant than that of the humble housefly who invades our kitchens, but this climate crusader really does look like a creature you'd rather swat away than welcome for dinner. At a mere 20 millimetres long, this fly hatches and matures in seven days, eating a mountain of food in comparison to its size along the way, up-cycling our leftovers, one magic maggot at a time. Hailing from Florida in the United States, these tropical superflies eat only when they are in their larvae stage, and if one homegrown Aussie start-up has its way, rather than buzz around the picnic blanket these insects will end up on the menu. Eat flies to save the planet, you ask? It's gut-heaving to think about it at first, but perhaps we need to shift our perspective.

Phoebe Gardner and Alex Arnold are wrestling with around 100 million of the critters every day in their fast-growing start-up, Bardee, which is a reference to the Aboriginal use of the word in various language groups: it means, you guessed it, edible insect. In their mid-twenties, the couple leads this climate-conscious venture that has grown from their North Melbourne living room to a fully-fledged pilot operation complete with a team of twenty people in two years. In stacked vats across eight labs housed at Sunshine in Melbourne's north-west, the soldier flies are offsetting 50 tonnes of carbon emissions every day by preventing wasted food sitting in landfill and emitting methane as it breaks down. 'Eating insects, and producing this new alternative type of protein might seem really new, but including insects in our diet and working in harmony with existing life cycles in Australia has been happening through really incredible agricultural technology developed by First Nations people over tens of thousands of years,' Gardner said. 'In many ways what we're doing is not new.'

The need to process our waste more efficiently is staggering: the 2021 Food Waste Index by the United Nations Environment Program report found an estimated 8–10 per cent of global greenhouse gas emissions are associated with food that is not consumed. In only a few days at Bardee HQ, food scraps from across Melbourne are transformed into two products: sterilised flies become a high-protein flour, an ingredient for pet food, and the manure is refined into an organic fertiliser that can be used for crops that has twice the nutritional profile of compost and 200 times the microbes, tiny organisms that support nutrients being absorbed by plants. 'We have entomologists from all around the world working in those labs to continue to breed the insects in that environment and to continue to optimise the nutritional profile of the protein that we're producing,' Gardner said. 'The only thing left after a week, which is just so incredible, is just the insects and the manure; there's nothing else.' What's more, trial packs of spicy dried flies are already a common snack food at Bardee's HQ, and a sumptuous sticky date cake, made with soldier fly oil, was on the menu for a colleague's birthday too.

How did Gardner, an architecture graduate running through early career jobs, and Arnold, an entomologist, come up with this idea? How did they think they could take on the global food waste system? Travelling for work, inspiration came from the garbage of Amsterdam, a place where the city's approach is enviable. There's no large wheelie bin collection each week in this city; folks bring their rubbish along on the morning commute, sorting waste into a bunch of common receptacles on the street. The pair started asking a question based on their growing professional expertise: what would it look like to recycle a whole city's waste, with insects? 'We thought this is going to be our biggest opportunity to make a real dent on climate change and prevent climate disaster, by having a big impact on the food system and how waste is recycled in cities, so we decided to give this relatively high-risk project a shot,' Gardner said. As soon as they returned to Melbourne, they made plans to go all-in on the idea. An accelerator program at Melbourne University helped kick things off with $20,000, and every cent went into their first science lab in a shipping container on campus.

In 2021 Bardee, raising $5 million in only two years, has investors buzzing. Plans are taking shape for a national operation, big thinking given that the start-up is diverting only 0.03 per cent of Melbourne's landfill so far. Gardner's quiet confidence and single-minded focus is remarkable for someone who is literally in her third post-university job. 'We're part of a movement to create a more circular economy, taking a waste stream that is currently under-utilised, taking what's there and adding value to it, recycling it back into the food system. So our vision is a world where Bardee is really successful and we maintain that 1.5-degree limit in total global warming.' Why did she think it could be done? 'This is a problem I care about and this is a solution, and us in combination, or in a small team, we just might be the best people in the world to tackle it. So, whatever comes in our way, we'll try to work it out, I think.' Their vision is one that could be the template for taking regenerative action: 'We see a future with lots of technologies that deal with these kinds of waste streams that currently aren't used. We think that lots of people need to give it their best shot at supporting and building these kinds of technologies because ultimately it will take lots of teams like ours to solve such a big and complex problem.'

TO MARKET

If you travel to the other end of the food start-up world you'll dig up Your Food Collective, a profit-for-purpose company that puts positive climate and health outcomes back on our plates. Founded in 2017 by cousins Lauren Branson and Cara Cooper, the start-up aims to change the way we select food by going straight to the source. 'Food is absolutely massive; it's at the intersection of so many different things that are important to our life, whether that's climate change and the environment, or just nourishing our bodies or connecting as communities,' said Branson, the organisation's CEO. 'We all eat three times a day, so food is one of the most powerful things that we can do as individuals to drive positive impact.' Branson, an ecologist with fifteen years in native species research and conservation behind her, moved her career to food because she found her own efforts to source fresh, locally grown food for her family near impossible, and climate change was a passion that demanded she make an outsized, systemic contribution.

> We are so far removed from what we need to be eating, because hundreds of years of policy has resulted in our food system ending up in this heavily commodified state. If you pare it all back and think about the essence of food, the whole reason for it is to nourish ourselves and to grow and be healthy, but that's not even what we're doing at the moment: 85 per cent of what's on the shelves in supermarkets is bad for us.

Your Food Collective's promise to customers is to source 95 per cent of its fresh fruit and vegetables from within 250 kilometres of the consumer's front door, with a commitment to transparency that has meant building deep relationships with a growing network of suppliers. 'We now work with over 100 local growers and producers; we have over 700 seasonal products that we work with and we've served over 20,000 customers, which is really exciting,' Branson said. 'We're focused on reaching as many people as we can and helping drive as many producers and farmers as we can to new farming methods that will regenerate our way of life through food.' Plans for the Newcastle-based start-up are huge: in five years Branson and Cooper want to reach $30 million in annual revenue and use the forensic knowledge built through managing a complicated supply chain to build a technology-enabled logistics business that will help people get a clear line of sight on our food choices. Think food miles and regenerative ratings printed on your weekly docket: from Branson's perspective it's this information that will transform the way we make our consumer choices.

Branson said the fundamental career shift while raising her young family has been full of uncertainty and a bundle of challenges: business as usual for start-ups comes with an enormous and relentless workload. So why did she do it? 'I didn't want to wake up when I was 50 or 60 or 70 thinking maybe I should have tried. At the end of the day, you're better off giving it a shot and becoming comfortable with being uncomfortable.'

Ahead of the United Nations Climate Change Conference (COP26) in Glasgow in late 2021, impact intelligence platform Holon IQ mapped the most promising 1000 start-ups in climate tech around the world, companies that were generally less than ten years old. The top 100 in Australia and New Zealand were captured in the list (Figure 7.1 gives an industry breakdown of the list), with a plethora of movers and shakers who are changing our transport, energy and food sectors right under our noses. The agrifood category features Vow Food, which is producing lab-grown meat that's attracted millions in venture capital, while Fable is already taking on the world with its vegan (mushroom-based) meat-substitute products, partnering with industry luminaries like Heston Blumenthal and launching its vegan burger partnership with fast-food franchise Grill'd. There's FloodMap, a flood forecasting and mapping platform, and Cecil, which helps people operate, track and report on nature-based projects that enhance ecosystems, preserve biodiversity and accelerate the transition to net zero. It's a hive of transformational potential.

SEAWEED SOLUTION

Almost every day of the week Dr Julia Reisser swims in the ocean, meditating in the place that has become a second home and her life's work. 'I've loved it since I was a kid,' she says from

FIGURE 7.1 EMERGING CLIMATE TECH IN AUSTRALIA AND NEW ZEALAND, 2021

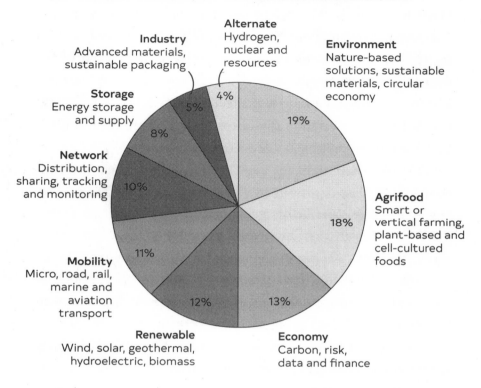

Source: holoniq.com

her home base at Cottesloe Beach, on Noongar Country, Western Australia. 'I feel very connected to the ocean, it is something that calms me down.' A marine scientist by training, Reisser worked as a cleaner when she arrived on Australia's west coast from her home in Brazil until securing a PhD scholarship and immersing herself in one of our modern world's greatest scourges: micro-plastics pollution. The more she studied and educated others on the scale and nature of the plastics pollution crisis, the more steadily she drifted to depression. 'We are an intelligent species. I do believe in the goodness of people at an individual level and we do care about our kids, but it does need radical change, and it needs to be quick. We need to incentivise our communities and our governments and our businesses to change.' The only option for Reisser was to work on solutions to help lift her out of the depths, first through advising early stage ocean clean-up ventures and then investors looking to make a difference. It was this work that led to the realisation that a compelling alternative to fossil fuel plastics had not been discovered.

That light-bulb moment quickly evolved into Uluu, a world-first plastic that is grown in vats using microbes feeding on seaweed in salt water before being processed into the base plastic material that can then be turned into almost anything: packaging, furniture, car parts, you name it. 'Most of our bio-plastics nowadays are made from terrestrial crops, and if you look at the fertilisers and the water and all the inputs that you need you end up still with a big carbon footprint, and also most of the bio-plastics don't compost.' Reisser's product, made from biodegradable polymer PHAs or polyhydroxyalkanoates, promises to absorb up to twenty times more CO_2 than reforestation, is grown three times faster than processes using land-based plant feedstock, reverses acidification and eutrophication (the increase in the concentration of plant nutrients like phosphorus and nitrogen in the water) and can biodegrade, in small pieces, on land and crucially in sea water too.

In under two years, Reisser and co-founder Michael Kingsbury built a team of microbiologists, biochemical engineers, marine scientists and business folks, raised $1.8 million, and transformed this kernel of an idea into the first proof of the concept, producing small flexible plastic discs and films. 'We take the carbon out of the oceans and then we can put it back into the environment as a fertiliser and create this beautiful circular bio-economy—that's our dream,' Reisser said. 'What it means is that our future wouldn't need petrochemicals anymore.' Seaweed is food for the polymer-producing microbes, sourced in Indonesia and sent to their demonstration plant at the University of Western Australia. Soon the supply chain will be pumping here too: according to the Australian Seaweed Institute, the aquatic industry could be worth $100 million by 2025, creating 1200 direct jobs in coastal communities and reducing domestic emissions by 3 per cent.

Reisser has seen up close the damage plastics cause to our oceans, but maintains that we will need plastics to make the world work, with alternatives like paper or aluminium bringing enormous carbon loads with them. I'm reminded of a revealing conversation I had with disability rights activist El Gibbs who rightly challenged me (and all greenies) on how campaigns to ban plastic straws have serious justice and equity considerations for people living with disability. Plastic is needed in multiple contexts, for people with disability and for medical uses, and the thought that we could make these products renewable in the truest sense is tantalising. I begin to imagine a waste-disposal machine that chops up your bio-plastic straws and shopping bags with your food scraps, all to go in the backyard compost bin to break down in days: could it be possible? Even with Reisser's qualifications, skilled team and bubbling enthusiasm, it all feels too far out to become a reality. Reisser maintains her confidence that the product could be in large-scale production within years, not decades. 'What I'm learning is to stop being afraid of taking the risk of doing my own thing,' Reisser said. 'And now that it can actually work, it's a very powerful tool to bring rapid change at the pace we need.'

What's going to help solutions like Bardee, Your Food Collective, Uluu and dozens more local climate start-ups take shape is a thriving ecosystem, where knowledge is shared across sectors, wins celebrated and failures allowed to happen, fast. 'Charles Darwin, for instance, I really respect him,' Reisser said. 'But I think that even more important than competition when it comes to survival is symbiosis; it's kind of being able to build and thrive in a system with many parts, and that comes down to communication and compromise and understanding.' There's that word again: symbiosis.

COLLABORATION CENTRAL

The lift doors open and I'm swept from the early summer evening rain into a cloud of home-grown climate tech start-ups in Sydney's CBD on Gadigal country. Logos with tantalising names like Novalith, GoTerra, Energy Synapse, Clean Our Blue, Sicona, Greener, baresop and Trace are peppering the packed room. Guests are invited to sample organic strawberries and chocolate sourced by Your Food Collective and stuffed into recycled cardboard cones, and packs of Circle Harvest's saltbush and rosemary corn chips made from crickets: converting insects to snack food is yesterday's breakthrough here at innovation HQ. Climate change often feels like an exercise in compromises we don't want to have to make, despite the ecological and moral imperative that we make them. But wandering around the crowded room I quickly see the sheer variety of new ways of ordering our world that are emerging in front of our eyes: even if only half of these ideas work, we can only be better off. Emerging from a couple of pandemic lockdowns, Climate Salad's inaugural get-together in late 2021 is twittering with the characteristic notes of networking: the crowd gets so rowdy that Mick Liubinskas, the network's instigator, repeatedly pleads with the people to keep it down while founders max out their 30-second slots pitching world-saving businesses to investors hungry for solutions with impact.

Liubinskas has an impressive background in technology, but these days is a climate-tech whisperer, advising and supporting founders and building connections of the human kind between tomorrow's business leaders and investors. Around $1 billion in investment dollars is represented across the 40 members of Climate Salad's investor group, including Main Sequence, Blackbird, Grok Ventures and Square Peg, along with seed investors like Investible, all keen to help climate-focused founders make their ideas a profitable reality. The launch event is packed with a potent combination of optimism and energy, and in a couple of hours my cup is full: there are just so many options within our grasp right now to address this massive challenge. Maybe it's the effect of my first major event after another big pandemic lockdown, but when I make it home that night, I'm awake for hours, my brain swirling with possibilities and potential.

Liubinskas is an adviser, investor and network builder who has supported thousands of successful start-ups through his career in start-up accelerators Pollenizer, Startmate, muru-D and EnergyLab in Australia and as Austrade's entrepreneur-in-residence in Silicon Valley. 'I've always been very environmentally driven but very personal about it, like it's my own one thousand small actions,' Liubinskas said of his personal sustainability journey. But this changed big time in 2016 when Trump landed in the White House; soon after, the Paris agreement was torn to shreds by the very nation that was pivotal to securing the agreement in the first place. Liubinskas gave his best friend in the world, Nathan Fabian (who these days chairs the European Platform on Sustainable Finance), some grief over the debacle. 'I said "hey, your job has got harder", and he just hit back at me twice as hard, messaging me with "yes, maybe it's time you got in the game".' Liubinskas took the words to heart, and decided to roll up his sleeves and use his skills, honed in the tech start-up and investing communities, to clean up the planet. 'I'm not a scientist; I'm not a biologist or a physicist, but I reached out to climate tech companies and they had a lot of the same problems as normal tech companies and I just started helping.'

Returning to Australia from the United States in 2019, Liubinskas found the ingredients for a thriving Australian climate tech community were there, they just needed to be combined. Liubinskas believes there are ten elements required to avert climate disaster: belief, collaboration, science, innovation, entrepreneurship, investment, consumers, corporates, governments and, without doubt, urgency. I can't help but think about the qualities we need to nurture, what I could see in plain sight at the Climate Salad event: curiosity, creativity, passion, wonder, resilience, empathy and a pretty great sense of humour, which is really important if you're going to try to start something no one's ever tried before. 'I think partly because of the beautiful combination of purpose and challenge, there's no one being defensive or blocking,' Liubinskas said of the emerging community. 'Collaboration levels are really high and it's really positive.'

Liubinskas described a recent exponential growth of interest in climate tech that's arrived at our shores, making the area a side interest no longer. 'I wish we did it twenty years ago, but like the old adage about the trees: the best time to plant them is twenty years ago, the second-best time to plant them is today.' There is a new clarity, Liubinskas said, that many traditional industries are going to struggle as the world shifts to cleaner energy and production, with coal, oil and gas, and plastics particularly under pressure. This trend is matched with another: a fast-growing realisation that the world needs to shift nearly everything we're doing. 'There's always been a sense that it's going to be minor, it's going to happen later, or it's going to happen to other people or we'll deal with it,' he said. 'But I think we're really starting to get it, and enough smart people have done that over the last year and there's been a positive tipping point. It's certainly not wholesale, it's not 90 per cent, but we don't need it to be 90 per cent. We need enough good people on the side of policy, finance, consumerism, innovation and entrepreneurship, and that's really tipped over.'

GIANT LEAP

One Small Step, an app that uses behavioural science to help people successfully adopt greener habits and reduce their carbon footprint, made it into the economy category in Holon IQ Australia–New Zealand top 100 climate tech start-ups in 2021. Founder Lily Dempster, a self-described climate activist and behavioural economics nerd, is a living example of why there is no such thing as a typical type in the emerging climate tech start-up ecosystem. You'd expect to encounter Dempster working in the halls of government or among the crowds in grassroots advocacy, and in earlier days she did both, working in the federal Department of Prime Minister and Cabinet before campaigning at GetUp to save the Renewable Energy Target and stop coal-seam gas projects. A background studying complex systems theory, behavioural economics and microeconomics mixed with politics, law and renewable energy came together when she delivered one of Australia's most successful climate-positive consumer-switching campaigns, when 35,000 people switched away from Australia's number-one carbon polluter, AGL, to Powershop. (In 2021, when Powershop was acquired by Shell, the consumer backlash was ferocious, with thousands switching to newer, cleaner start-up energy retailers like Amber, which also made it onto Holon IQ's list.) Second-year university was the turning point for Dempster, when she watched Al Gore's global filmic phenomenon *An Inconvenient Truth* and life was never the same again. 'I was at university because I cared a lot about social justice,' Dempster said. 'The inequitable impacts of climate change I found horrifying, both intergenerationally and for the people who have the least responsibility for causing the problem: the most vulnerable people in our communities, internationally and here, are going to be most badly affected. So that's why I decided to, honestly, dedicate my life to it, and I've been doing that and it's been challenging and purposeful.'

Running the consumer campaigns at GetUp, Dempster saw the impact on emissions and on markets from a few thousand people taking action together and came to the realisation that pushing for top-down changes through policy and law reform was critical, but so were the bottom-up actions that people take in daily lives. 'When you look at consumption-based carbon accounting, the influence that consumers can have when they collectivise is pretty significant, especially people in developed economies, where you have really high personal footprints.' One Small Step, a social enterprise, launched in 2020 and a year later its tailored programs across energy, waste, purchases, food and transport had saved 3500 tonnes of greenhouse gas emissions thanks to the efforts of 18,000 initial users. If 50,000 people reduced their carbon footprints by one-quarter over twelve months, Dempster said, 250,000 tonnes of carbon emissions would be saved (based on an average of 20 tonnes of CO_2 equivalent as a baseline), roughly the same amount as planting 4 million tree seedlings or taking 54,000 cars off the road.

A big part of One Small Step's approach is trying to encourage positive spillover effects, a concept in the field of behavioural economics that suggests that one pro-environmental behaviour leads to another. One of the dangerous things in climate circles is judging people who aren't 'doing enough' but research shows this is a misplaced way to think about behavioural change. Studies over several years demonstrate that encouraging 'private' behaviours, such as consumer purchasing and waste behaviours, increase people's engagement in 'public' behaviours that can help to influence policy, such as signing petitions or going to protest rallies. Research has also found self-identity influences our behaviours, and performing those behaviours connected with our own identity can reinforce that identity too. The day-to-day, sometimes entirely mundane, habits of climate action, things like taking your own shopping bags to the supermarket and choosing to ride or walk rather than drive, have two areas of benefit when it comes to how we behave: first, the scale of everyone doing these can make a tangible difference to Australia's collective carbon footprint, and, second, these actions help to reinforce our identity as personal climate action powerhouses who might then do more in all spheres of private and public life.

Some useful research by behavioural experts from the University of Queensland and Monash University in 2017 confirms this idea. Nita Lauren, Liam Smith, Winnifred Louis and Angela Dean found that when we are reminded of what we've done to help the environment in the past, we feel encouraged to keep up our climate-positive behaviours. What's more, we are more likely to spread our wings and take climate-positive actions in public. Did someone once say something about big things growing from little things? That's another way to think of this spillover effect, and why it's important to encourage more action from individuals.

None of this is a surprise to Dempster. 'We know some behaviours are socially contagious, so we can shift social norms more effectively and quickly with a bottom-up grassroots approach, and we can shift corporate behaviour effectively by getting people to shift consumer demands, by supporting the growth of zero-carbon secular economy businesses.' The aim is to get people to reduce their individual carbon footprints to 2 tonnes of CO_2 or below, which is in line with the United Nations' 2050 goal. When I chat to Dempster, an app feature has been added to One Small Step so users can construct team challenges, while launches are around the corner in the United States and United Kingdom. Their next stop? With expert advice from the former director of community at online global phenomenon Reddit, the app is soon to launch features that will allow users to see what like-minded people and organisations are doing in their local area, and allow them to jump in. 'There's a need for a sense of togetherness, and that needs to be real; caring about where you live, the people around you, is part of that,' Dempster said.

One Small Step doesn't outsource climate solutions: just like a fitness or personal finance program, you've got to be doing the work to see the changes happen. How does it help if you've got to DIY everything? The brain science at work here, Dempster explained, is reducing the cognitive load in our lives, while also providing a sense of agency. Completing the app's five-minute quiz, I was dismayed to see that despite the masses of solar panels on our rooftop, reduced meat eating and Covid-inspired light car use and almost zero flying, I'm at 4.5 tonnes of CO_2, well above the goal of 2 tonnes. My brain yells 'FAIL' but, given my self-identity depends on it, I persist. The app conjures up a personalised sustainability plan with a few quick wins on cutting my car's carbon, going carbon neutral for electricity and making changes to reduce waste. New habits are something I'm generally terrible at, unless it's a 5 p.m. glass of unwind wine or procrastinating on tasks with a good Netflix binge-watch, but after a couple of months I found myself sorting my soft plastics and hanging wet plastic bags on the line to dry like I'd been doing it for decades. Habits, it seems, can come creeping in until the day when we don't give them a second thought.

CATALYSE THIS

Climate impacts are personal for Usman Iftikhar, co-founder and CEO of Catalysr, an award-winning pre-accelerator that supports immigrant and refugee entrepreneurs. In 2010 while studying engineering at university and living at home with his parents in Pakistan's north-western Khyber Pakhtunkhwa province, he was caught in one of the worst floods in the country's history. 'It was raining so much that in the apartment complex we used to live in, the backyards filled up with water, it was overflowing,' Usman said. 'It was five or six feet [1.5 to 1.8 metres] of water, all our cars were submerged and there was a point when the whole boundary wall of the complex literally fell down.' On the ground floor of their apartment, the rising water began flowing through the windows, and Usman rushed outside to try to close them, but the water resisted all efforts; the windows broke under the strain and the falling sheets of glass sliced his arm, where a hefty scar still reminds him of that day. The one-in-100-year flood was described by the secretary-general of the United Nations, Ban Ki-moon, as the worst disaster he had ever seen, with 2000 dead and approximately 20 million people directly affected by the catastrophe. 'I'm very lucky that my parents are doctors; they patched me up but we couldn't get to the hospital. I lost a lot of blood. It's a very vivid memory.'

Usman harnessed this and many more experiences in Pakistan and Australia to kick off Catalysr, a start-up that helps other start-ups get on their feet. Launching in 2016 in Australia, more than 520 founders have created 175-plus businesses with Catalysr's support, including climate-connected organisations like Natural Panaa, a social enterprise that sells palm leaf products to replace disposable plastic tableware, and Spiral Blue, which is building the next generation of Earth-observation satellites. A community of advisers, investors and mentors has bloomed to around 1200 people and only eighteen months post-launch, Usman was named the Commonwealth's Young Person of the Year, receiving the gong from Prince Harry himself. Usman agrees that climate and sustainability start-ups will be the next big tech boom, which will bring many out-there ideas. 'I think the interesting challenge is going to be how we separate the useful solutions from the hype and the crazy capitalisations that are happening for start-ups. Even if the idea is well intentioned, it doesn't mean that a lot of solutions are good solutions or are worth scaling up, because they might lock us into positions which may not be great for the longer term. That is why I think a systems approach is so important when thinking about climate solutions.'

Arriving in Australia in 2013 as an international student, Usman completed Al Gore's Climate Reality training and volunteered with the Australian Youth Climate Coalition, where he learned how climate change will cause increasingly frequent and intense natural disasters like the Pakistan floods. These days he is spending more time reflecting on how entrepreneurship and global networks of businesses can help work on solutions for how people will move around the world, either by choice or because they are forced to leave their homelands as a result of climate change. It's likely to be an enormous challenge, when you consider that rising seas alone could result in 1.4 billion refugees by 2060, according to research by Cornell University. But rather than get overwhelmed by these challenges, Usman recommended bringing an entrepreneurial approach, which essentially means falling in love with the problem. 'One of the key mindsets to have is curiosity, and not to just go with what you've been told,' he said. 'Experiment, be curious and try to find the best possible solutions.'

TOGETHER WE CAN . . . *make it*

✳ When it comes to climate change, we need to fall in love with the problem, not fear it, if we're going to spark the curiosity and creativity we need to find the best solutions.

✳ Changing our habits can start with a small step, but it won't end there, because climate-positive behaviours are more likely to spill over to other behaviours, not only in your personal life but in the wider community too.

✳ Building supportive networks is key to learning, testing and experimenting with solutions. Collaboration turns a spark of an idea into reality.

FURTHER READING

✳ Bardee, n.d., viewed 20 February 2022:
   https://bardee.com

✳ Catalysr, 2018–, viewed 20 February 2022:
   https://catalysr.com.au

✳ Geisler, Charles & Currens, Ben, 'Impediments to inland resettlement under conditions of accelerated sea level rise', *Land Use Policy*, vol. 66, July 2017, pp. 322–30:
   www.sciencedirect.com/science/article/abs/pii/S0264837715301812

✳ Holon IQ, *Global Climate Tech Landscape 1.0*, 26 October 2021, viewed 20 February 2022:
www.globalclimatelandscape.org

✳ Lauren, Nita, Smith, Liam, Lewis, Winnifred & Dean, Angela, 'Promoting spillover: How past behaviours increase environmental intentions by cueing self-perceptions', *Environment and Behaviour*, vol. 50, no. 3, April 2019:
https://doi.org/10.1177/0013916517740408

✳ Liubinskas, Mick, 'Climate action: The power of ten', *Climate Salad*, 29 November 2021, viewed 20 February 2022:
www.climatesalad.com/posts/climate-action-the-power-of-ten

✳ Natural Panaa, 2021–, viewed 20 February 2022:
https://naturalpanaa.com

✳ One Small Step, 2021–, viewed 20 February 2022:
www.onesmallstepapp.com

✳ Spiral Blue, 2021–, viewed 20 February 2022:
www.spiralblue.space

✳ Uluu, n.d., viewed 20 February 2022:
www.uluu.com.au

✳ United Nations Environment Program, *Food Waste Index Report 2021*, viewed 18 March 2022:
https://wedocs.unep.org/bitstream/handle/20.500.11822/35280/FoodWaste.pdf

✳ Your Food Collective, 2022, viewed 20 February 2022:
https://yourfoodcollective.com

CHAPTER 8

ESCALATING OUR ACTION

'If we want hope, we have to earn it!'

—CHLOE, GROUNDSWELL GIVING MEMBER

If you're on the outside looking into the world of global finance and investment banking, it can feel like some kind of alternative universe. Beings from other worlds with an impenetrable language somehow make the whole global financial system work—and if the system fails, we're all apparently stuffed. Money markets are characterised in our culture as a superhuman life form with an insatiable appetite and fluctuating moods to boot, but in reality they are just another complex system that's integrated with many others, managed like any other system in our society by people with a particular set of skills. It's an area of expertise I find as obscure as the intricacies of climate science, but nonetheless the individuals who are neck-deep in global finance are simply people too, with emotions, dreams, hopes and fears, the lot. And just like you and me, they can't hide from the urgency of impending climate catastrophe and are responding using the ratchets in their specialised toolkits.

I chatted to Matt Nacard, Nathan Parkin and Robyn Parkin of Ethical Partners, a new-ish investment fund that knows how to make money do better for the world: Nacard and Nathan Parkin are co-founders and Robyn Parkin is head of sustainability at the fund. After more than twenty years in various roles in financial markets including at Macquarie Bank, in 2017 the co-founders decided to start their own niche investment company, which, with more than $2.5 billion under management, gives a whole new meaning to 'boutique'. Over a couple of hours I received a crash course in how investing for climate's sake is scaling up. 'There's a movement of capital that's occurring here in Australia and around the world and engagement of members that hasn't happened before,' says Nathan Parkin, who is also the firm's investment director. 'We also see capital moving faster than governments . . . it's quite exciting to place ourselves in what was a niche in the market but is now becoming simply the way you manage money.' Clients include superannuation funds, wealthy individuals and family offices, charities and schools, all of whom want to see a decent return on investment and ethical considerations across climate change, human rights, corruption and more in the one package. How does it work? 'We assess the stocks under a range of financial criteria as well as ethical considerations, and then form a view on a company,' Nacard said. 'We then put a valuation on that company and decide how much of it we want to own on behalf of our clients. Engagement with the company is also crucial because we believe this is beneficial to both the company's future performance and the world.'

Ethical investment is surging, with 30 per cent growth in 2020 to $1.28 trillion under management, according to the Responsible

Investment Association Australasia's 2021 annual benchmarking report, and climate change was the leading sustainability theme of the report's survey of 28 respondents who offer sustainability-themed investment products (Figure 8.1 shows the full range of sustainable investments). The potential is striking, not only for investors, but for the rest of us: Australia and New Zealand's Investor Group on Climate Change found that if our nation adopts a 2030 emissions goal in line with the Paris agreement, combined with a net zero target by 2050 and supporting policies, we could unlock $131 billion in investment and job opportunities by the end of the decade, with investment set to benefit regional Australia. We've still got a way to go: Ethical Partners' 2021 survey of 216 ASX-listed companies found that only 60 per cent of companies clearly disclose emissions data to shareholders; only half of those surveyed have set any emissions targets; and a mere 30 per cent have a net zero target or report fully in line with measures recommended by the Taskforce on Climate-related Financial Disclosures. 'Net zero targets are powerful, not least because they are being adopted by governments around the world, including most of Australia's key trading partners,' Nacard said. 'But when companies codify targets into their business and operations, it means the calculation of the company's long-term value is transformed. All of a sudden the intrinsic value of a lot of these companies changes significantly, which could be viewed as a hit to the bottom line in perpetuity, but it presents opportunities for companies that leap in, and many are.'

Ethical Partners' investor members sit in two broad categories: those keen to minimise risk to investments, and those who make doing the right thing with their money their top priority. 'In some places, our clients are becoming experts on these issues and it's moving more quickly than in any trend I've observed in at least

FIGURE 8.1 SUSTAINABILITY-THEMED INVESTMENTS, BY THEME (PERCENTAGE OF ASSETS UNDER MANAGEMENT), IN 2020 AND 2019

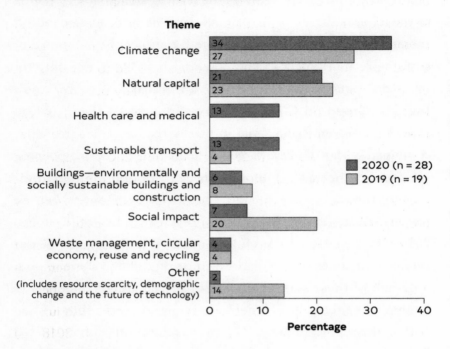

Theme

Source: RIAA, *Responsible Investment Benchmark Report*, 2021

20 to 25 years in financial markets,' Nathan Parkin said. On the risk minimisation side, investors are keen to ensure the companies they support don't fall foul of regulators or end up with stranded assets because of changing policies at global or national levels or because of shifting consumer demand. Investors who prioritise ethical considerations rely on the knowledge base that Ethical Partners has built, with some 600 data points informing the firm's recommendations on the environmental, social and governance performance of companies, or ESG. 'It's a lot of work to get those independent and varied sources of data, but we think that it is crucial to get a better quality and range of information to make

our decisions,' said Robyn Parkin. Investing on an ethical basis also delivers another critical outcome: influence. 'We are a substantial shareholder in many companies: we might hold 7 or 8 per cent of a company,' Nacard said. 'That means that the chair of the board, board members and CEO will want to and need to engage with us. We're not an activist public investor typically but they know in the back of their minds that we vote, so engagement is a very important part of the process.'

The corporate world is increasingly building ESG principles into strategy and operations, after years of agitation from investors, consumers and shareholders in Australia, including work by homegrown activist groups like Market Forces and the Australasian Centre for Corporate Responsibility, great organisations to look at if you're thinking about how to shift your banking or superannuation to be climate-positive performers. Analysis by Bloomberg in early 2021 found ESG-connected assets under management jumped from $22.8 trillion globally in 2016 to $30.6 trillion in 2018 and could climb to $53 trillion by 2025, more than one-third of the projected $140.5 trillion global total. However, using ESG principles is an approach that is complicated by hundreds of sets of definitions and measurement standards applied by individual companies and investors around the world. The common criticism of the 'E' in ESG is that it's scrubbing clean the perception of sub-standard corporate climate and environmental performance. This is all set to change, with the International Financial Reporting Standards Foundation having formed a new International Sustainability Standards Board in November 2021. The new board will develop a baseline of high-quality sustainability disclosure standards so, in time, investors will have the ability to compare companies on a like-for-like basis. There's a bunch of resistance now, just as there was when global accounting standards were introduced two decades ago, but soon there will be nowhere to hide for companies that attempt to greenwash their climate credentials.

US investment firm Pimco found that, before 2018, less than 1 per cent of corporate earnings calls across 10,000 global companies mentioned ESG principles; this rose to 5 per cent in 2019 but leapt to 19 per cent of calls analysed in the May 2021 reporting season. In Australia, the changes are substantial. An Ethical Partners internal analysis of every 2020–21 full-year financial results presentation for all S&P/ASX 300 companies that reported in the 2021 season found mentions of ESG at 61 per cent. Outcomes matter most, of course, but the changes in conversations in reporting seasons (the moment each year when companies are put under the spotlight by shareholders and fund managers), are a key indicator of the increasing understanding that ESG risks and opportunities are integral parts of a company's financial operations, accounting and strategy—and of an investor's analysis. The number of corporate disclosures for climate and environmental impacts are steadily growing, according to not-for-profit CDP, an organisation that runs a global disclosure system for investors, companies, cities, states and regions. Back in 2003 a few hundred companies around the world were disclosing climate performance, as requested by investors, purchasers and city stakeholders; this rose to more than 13,000 in 2021. These can be seen in Figure 8.2.

For Nacard, it's about using the resources and skills you have to make an impact. 'Am I out on the street with a sign when we have the marches? No, I'm not,' he said. 'Why am I not out there? It's because I think I can actually use my skills better doing what we're doing here. That's not to say that those protesting about key areas of concern such as climate change aren't doing the right thing, not at all. I think each individual needs to assess how they can make the biggest and most significant difference.' And that's

FIGURE 8.2 GROWTH IN CLIMATE CHANGE DISCLOSURE FROM 2003 TO 2021

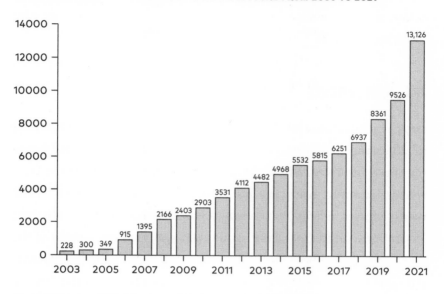

Source: Based on CDP, 'The A List 2021', www.cdp.net/en/companies/
companies-scores

where you come in, whether you have megabucks in the bank or not. 'We've got to get back to the first principles of what's driving this, what's right at the beginning of it all,' Nacard said, 'and I think it's the attitude of the person in the street, the average person in Australia.' Most of us don't focus on investment every day, with weekly mortgage repayments or rent, the quarterly bills and all the other elements of running our lives preoccupying our minds. But back in the early 1990s, when I started my first job on the checkout at the local supermarket, compulsory super began adding up. Superannuation means that every employed Australian is an investor, and these days $3.3 trillion total funds is under management, now the third-largest pool of funds in the world. Yes, friend, most of us are investors on a global scale through the 9–5 grind, but once you've switched your super and your banking to climate-friendly options, how else can we make time in the office climate-positive? It's time to make our working day also work for climate.

PEAK PERFORMANCE

Lucy Piper, founder of Work For Climate, spent a decade working in advertising, communications and branding work that led her to the travel industry, where she had a globetrotting role with ethical adventure travel company Intrepid. 'I was travelling around the world filming videos and basically having what I felt was the best job in the world,' Piper said. Stints telling stories from some of the planet's most remote and seductive locations coupled with travelling annually from her newish Australian home to the United Kingdom to see family all added up to a carbon footprint Piper said she is fairly ashamed of these days. But what this experience did deliver was an education in causes and solutions of climate change and, given the industry's large emissions profile, how business can play an active role in climate mitigation. It was the warm-up for her main act: when her son was born, Piper reassessed her priorities. 'I can't remember where I've heard it quoted, but someone once said that once you become a mother, you become a mother to every child, and I felt that really deeply,' Piper said, realising her complicity in the problem, and her responsibility for playing a role fixing it. 'I was a part of the problem because I wasn't changing my lifestyle and I wasn't using my influence,' she said. 'That moment of realisation felt heavy, really heavy, and I felt like an adult probably for the first time.'

When my son asks me when he's older, 'Mum, what were you doing during that decade, when there was still an opportunity for us to mitigate the worst impacts of the climate crisis? What were you doing?' and I just feel like I have to be able to tell him that I did everything that was humanly possible, everything that I had the energy for. I didn't know anything about climate science and when I came into this space, I knew nothing about taking action on climate. But all I know is that I have passion and energy, and I just wanted to apply that to getting other people to apply their passion and energy too.

Piper made a life-changing decision to escalate her action by going all-in through her work, a choice people are making across the country as the nature and scale of the climate crisis comes knocking with a severe summer's day or creeping changes to our homelands. Considering the practical skills built over the course of her career as a marketer and communicator, Piper also reflected on the experiences she'd had influencing Intrepid's management around the organisation's gender equity and diversity and inclusion goals. 'I learned how influence works inside a corporation and the different levers you have available to get a business on board.'

Piper was drawn to a new start-up, Work For Climate, which helps employees of corporations across Australia influence senior leaders to ratchet up climate commitments that have the potential to shift emissions across multiple sectors of the economy. 'Work For Climate exists to help individuals take action: people who are feeling frustrated, who are feeling disempowered, who feel like their company is paying lip service and greenwashing, who want to do more but don't know how,' Piper said. 'We help those people become the change-maker they want to be inside their company.' Given that, according to the Organisation for Economic Co-operation and Development, we spend close to 1700 hours annually at work, escalating action in this arena makes good sense.

Work For Climate zeroes in on four areas significant to reducing a company's carbon footprint: switching to 100 per cent renewable energy sources; committing to 50 per cent emissions reduction by 2030; ensuring companies advocate for improved climate policy from governments; and ensuring their supply chains are carbon neutral. Work For Climate's latest playbook of information helps employees encourage companies to switch default superannuation funds, the one you're automatically added to when you join a new employer, to climate-friendly providers.

The initiative comes at a good time, as some of Australia's biggest companies begin to feel the heat: campaigns from Greenpeace and RE100 have seen corporate players move substantial chunks of electricity demand to renewables. Between 2018 and 2021 around 10,000 gigawatt hours of annual electricity use has been committed to renewables (based on 2019–20 levels), driving interest in clean energy contracts. The Business Renewables Centre Australia's 2021 *State of the Market* report found corporate power-purchasing agreements were a source of renewable energy demand in challenging conditions, with contracts signed for 1.3 gigawatts of clean energy capacity (roughly the same capacity as a mid-sized coal-fired power station). The report also found buyers of clean energy are motivated by sustainability ahead of cost savings. 'It's a moment in time rather than unifying characteristics of our cohort participants; they come from very different ages, different professions and different experiences,' Piper said. 'Everyone in our cohorts has had enough.'

COLLABORATION CENTRAL

Investors and employees are pooling their efforts, so what's the deal with giving away money to help the cause? ClimateWorks

Global Intelligence estimated that total global philanthropic giving by foundations and individuals grew to US$750 billion in 2020, up 14 per cent on the previous year. These kinds of dollars sound like we should have solved everything already, but it's shocking to learn that these climate-focused billions add up to less than 2 per cent of total giving around the world. Here in Australia, environmental charities received $164 million annually in donations and bequests reported to Australian Charities and Not-for-profits Commission (ACNC) in 2019: an encouraging increase of 10.4 per cent on the previous year, but pretty minuscule compared the $11.8 billion in donations received by charities in the same reporting period. Total revenue of environmental charities was A$862 million in 2019, a mere 0.5 per cent of total charities revenue in Australia, according to an analysis of ACNC data conducted by the Australian Environmental Grantmakers Network and Groundswell Giving (see Figure 8.3). It's a drop in the ocean when you compare the revenue of the top six climate-focused non-government organisations to that of the top climate-polluting companies: the job environmental charities do is pretty remarkable.

FIGURE 8.3 AUSTRALIAN REVENUE IN 2020 (A$ BILLION)

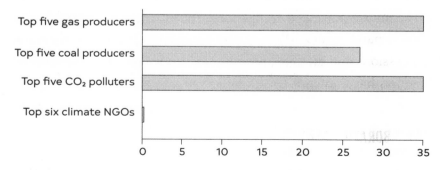

Source: Australian Environmental Grantmakers Network and Groundswell Giving, *A Rising Tide: Climate and Environmental Giving 2021*

Groundswell Giving, a newer NGO on the scene, is all about collaboration, community and participation in climate-focused philanthropy. Joining Groundswell starts with a $20 weekly donation, or $250 a quarter or $1000 a year, which means you're a member with the right to vote in quarterly funding rounds that decide which organisations receive pooled member funds given out in $40,000 chunks. 'We have a family that puts in their $1000 for the year and then decides where their family vote should go, and some of our members give up their Netflix subscription to sign up to Groundswell,' said co-founder and director Arielle Gamble. Organisations apply for grants and are shortlisted before each grant round, before members vote. Groundswell's grants have supported inspiring organisations like Martuwarra Fitzroy River Council, a powerful Aboriginal-led organisation that is campaigning to stop fracking for fossil gas in Western Australia. Bushfire Survivors for Climate Action's 2021 grant helped support successful legal action with the Environmental Defenders Office Australia that compels the New South Wales Environment Protection Agency to develop policies that measure and regulate greenhouse gases in the state. There's also Central Queensland Energy futures, which aims to bring together leaders across coal-dependent regions in the state to accelerate transition to a zero emissions future. Farmers For Climate Action, Seed Indigenous Youth Climate Network, the Climate and Health Alliance, and Environmental Justice Australia are also among the new alumni of Groundswell's grant recipients.

Groundswell aims to accelerate climate action in Australia by creating a community of new givers who fund strategic, high-impact climate advocacy. The mission developed after Gamble and co-founders Anna Rose and Clare Ainsworth-Herschell identified the sheer diversity and scale of advocacy efforts taking place across the country that needed to scale up quickly, and the number of people in their own networks who were unsure of how they could help, or where to direct their donations. 'There's all these people out there who are really worried but don't know how to step in or what to do,' Gamble said. 'They feel disempowered and over-whelmed by the scale of the crisis, and are seeking education and community.' Considering these two challenges, and after assessing giving circles in Australia and the United States, Groundswell was launched in late 2019, amid the Black Summer fires. 'We initially saw it as a much smaller giving circle purely for philanthropists but when we launched we were getting calls from all over the country, from Tasmania to Western Australia,' Gamble said. 'So we could see that it was so important to be as inclusive as we could in how we designed this giving platform.' In just under two years since launching, Groundswell has given away $800,000 and plans to do the same in the coming twelve months, granting twice the funds in half the time frame, and boasts 560 members including high-profile folks like award-winning chef Kylie Kwong and fashion designer Kit Willow, as well as hundreds of everyday people who are throwing in their $20 a week; the box 'Voices of Groundswell Giving' provides an insight into their motivation. Only 10 per cent of the membership are wealthy individuals. 'We know we need to scale up and deploy funds to match the scale of the crisis and that's dependent on new members; more members means more grants we can give out to climate action,' Gamble said.

A big part of Groundswell's work is to provide due diligence on grantees through its shortlisting process, saving time for members while dispelling myths about outsized climate action along the way. 'A lot of people I know wouldn't know much about what advocacy involves; there's just a lack of education and communication about what collective action and advocacy looks like,' Gamble said. 'So what we've really tried to do through

VOICES OF GROUNDSWELL GIVING

'I'm happy to be able to help support the great work you are doing! We are going to use all our unique gifts and voices to propel this idea to as many people as we possibly can in every way we can in order to get there. Game on.'

'I love that Groundswell makes it possible for everyday people to be philanthropists. We don't have to be individually wealthy—through pooling our resources collectively we can provide significant funding to the community organisers and activists whose work is so critical.'

'Why have I waited so long to put my money where my mouth is? Literally. The planet feeds me in every way, every day. If giving money to fight for it is what we need to do to make positive change then PLEASE TAKE MY MONEY.'

Groundswell's communications and social media is explain why advocacy is a really vital tool for shifting systems, and break down the misconception that you need to be a kind of green-haired leftie to be part of creating change.' Gamble, an artist, graphic designer and community organiser, initially thought her contribution to climate change could be creating a climate-focused arts project to travel around Australia. When she discovered how much activity was already taking place around Australia, she shifted her focus, deciding to help existing work scale up. In the short time since Groundswell landed on the giving scene, Gamble has observed people who are experiencing a similar journey of discovery. 'What we need is for philanthropy in Australia to scale up significantly,' she said. 'We all have our different superpowers and, for people with funds, the amount of change they can unlock is astronomical: right now there just isn't time for holding back.'

One collaborator in the emerging funding community is Stephen Pfeiffer, who inherited a few million in the late 2000s, something most of us spend weekends daydreaming about. Pfeiffer found having this amount of cash just landing in his lap so profoundly unsettling that he left it alone in an investment fund for thirteen-odd years, continuing with life as it was, working in his education-focused roles first as a high school history teacher and then in student recruitment and support in the university sector. 'It scared the bejesus out of me: I didn't have a clue what I was supposed to do with it at that point,' Pfeiffer said. But as the money sat there accumulating returns, Pfeiffer's life as usual was coming under self-scrutiny too, as his knowledge of the climate emergency developed. 'My work as a history teacher taught me that civilisations and cultures have faced existential threats before: invasion, colonisation, slavery, genocide, plagues, decline of resources, war and nuclear weapons,' he said. 'I've always been moved by anthropogenic climate change and would often link it to various parts of the syllabus the students were studying, and it's clear that what we face today may be the greatest existential threat that humanity has ever faced.' Pfeiffer came to the stark realisation that a total response was required, similar to those of the wartime economies of combatant nations in World War I and World War II: the mobilisation of all available people and resources to achieve the rapid transition to a carbon-zero global economy. 'It's incredibly alarming, reading about 2-degree tipping points and the Earth getting into a hothouse scenario I find very terrifying: I try not to think too long term in the sense of where those trajectories are at!' But the fear was motivating, and so at the ripe old age of 38, Pfeiffer realised the substantial resources he had access to could be put to work and his work as a philanthropist began. 'There's other things you'd really rather do with your life, but I'm compelled to do this; I can't look at myself in the mirror without being a part of the solution.'

Pfeiffer was inspired by the words of global philanthropic leader Larry Kramer, president of the William and Flora Hewlett Foundation and the former dean of Stanford Law School, who stressed the importance of climate-focused funding as the sector's number-one priority. 'Any grant maker who just chugs along on the same issues without addressing climate is, truly, fiddling while the world burns—particularly given the certainty that whatever short-term progress is made through these efforts will be lost if climate change continues unchecked,' Kramer wrote, and Pfeiffer agreed wholeheartedly. Pfeiffer's first steps were to join the Australian Environmental Grantmakers Network and Groundswell in 2020, to learn how homegrown philanthropists go about managing and giving money. This exposed him to scores of climate organisations focusing on different aspects of this colossal challenge. 'It made me realise that this is a complex issue and there is no silver bullet,' Pfeiffer said. 'It's more like a jigsaw puzzle where all these different organisations have a unique place and are working together and connecting with each other to create those solutions. No one can do this by themselves or in isolation from each other.' A year later, after switching investment managers and deciding to give away between one-quarter and one-third of his total bucket as soon as possible, the Stephen Pfeiffer Fund for a 100% Sustainable World (SPF 100) was born.

SPF 100 aims to accelerate the transition to a 100 per cent clean energy economy and environmentally sustainable society, where humanity can live and thrive within our ecological limits and the planetary boundaries, including our carbon budget. When

Pfeiffer and I speak, the fund is in the process of selecting five to six climate organisations focused on action, advocacy and solutions to receive three-year grants of about $70,000–$75,000 per year, which should mean organisations can hire a new staff member for three years, providing the income certainty that smaller climate charities are desperate for. A suite of one-off projects will also be supported. 'It's not a matter of "will we transition or not?" towards a carbon-free economy—that's absolutely going to happen, and we will absolutely feel the impact of climate change and that will increase,' Pfeiffer said. 'The only question is one of time: do we want to do this quickly and reduce the worst of it, or do we want to drag our heels and risk the very worst of it? That's how I've approached it.' Rather than decide to invest the money and make far smaller donations each year from the fund's returns, Pfeiffer chose to implement a 'spend down' approach to funding after first being introduced to this type of giving-strategy in conversations with Anna Rose. In the three years of the fund's life, all the money will be distributed, like an up-front investment in passionate people who are accelerating action in their day-to-day work. 'It's very stimulating work; it's very enriching. I've noticed a shift in my mood over the past twelve months, going from something close to depression, and anger and frustration to just feeling momentum, possibilities, ideas, creativity and inspiration from people who are working their backsides off.'

TOGETHER WE CAN . . . *maximise our impact*

* Finance and business are shifting to climate solutions, and the scale and pace of this shift is accelerating. Look to organisations that make it easier for you to work out where your money and work time can be directed.

* Think about how your time and money can become climate-positive. Start with resources you have at hand: the place you work, where your money is invested and the organisation you bank with.

* Leverage is best created through communities and groups, so consider how you can create outsized impact through the communities and companies you're already a part of, playing in, buying from or working for: your group efforts can be more than the sum of their parts.

FURTHER READING

* Australian Charities and Not-for-profits Commission, *Australian Charities Report—7th Edition*, 17 May 2021, viewed 20 February 2022:
www.acnc.gov.au/tools/reports/
australian-charities-report-7th-edition

* CDP, 'The A List 2021', CDP, n.d., viewed 20 February 2022:
www.cdp.net/en/companies/companies-scores

* Desanlis, Helene, Matsumae, Elin, Roeyer, Hannah, et al., *Funding Trends 2021: Climate change mitigation philanthropy*, report, ClimateWorks Foundation, 7 October 2021, viewed 20 February 2022:
www.climateworks.org/report/
funding-trends-2021-climate-change-mitigation-philanthropy

✳ Diab, Adeline & Martin Adams, Gina, 'ESG assets may hit $53 trillion by 2025, a third of global AUM', Bloomberg Professional Services, 23 February 2021, viewed 20 February 2022: www.bloomberg.com/professional/blog/ esg-assets-may-hit-53-trillion-by-2025-a-third-of-global-aum

✳ Groundswell Giving and Australian Environmental Grantmakers Network, *A Rising Tide: Climate and environmental giving 2021*, report, 28 June 2021, viewed 20 February 2022: www.groundswellgiving.org/2021-giving-report

✳ Investor Group on Climate Change, *Mapping Australia's Net Zero Investment Opportunity*, report, October 2020, viewed 20 February 2022: https://igcc.org.au/wp-content/uploads/2020/10/121020_IGCC-Report_Net-Zero-Investment-Opportunity.pdf

✳ Kramer, Larry, 'Philanthropy must stop fiddling while the world burns', *Chronicle of Philanthropy*, 7 January 2020, viewed 20 February 2022: www.philanthropy.com/article/ philanthropy-must-stop-fiddling-while-the-world-burns

✳ Responsible Investment Association Australasia, *Responsible Investment Benchmark Report: 2021 Australia*, September 2021, viewed 20 February 2022: https://responsibleinvestment.org/resources/benchmark-report

✳ Taskforce on Climate-related Financial Disclosures, 2022–, viewed 20 February 2022: www.fsb-tcfd.org

CHAPTER 9

CLEANING UP A DIRTY WORD

'It is past the time of sitting back, hoping that others will
do something. It is time to turn up, speak up and step up.
The outcomes are worth it. The nation needs it.'

—CATHY MCGOWAN AO

Anna Josephson, like so many of us, was frustrated with the
state of Australian politics, increasingly worried about the lack
of action on climate change and appalled by her local member
of parliament, the former prime minister and renowned budgie
smuggler–clad climate sceptic Tony Abbott. 'Climate was one of
the major things that really, really annoyed me,' she said. 'He tried
to get rid of the Climate Council, he killed the carbon tax and
all the good things that were starting to happen.' Josephson did
what many of us do when it comes to Australia's federal climate
policy debate: commiserated, with friends. 'There's a group of us
every Wednesday morning, we run and we swim and we have
breakfast, and so we complained,' Josephson said, laughing at the

memory. 'We developed this really good running pace, so you could run and complain at the same time.' Josephson, who in the 1980s moved from Sweden to Australia, settling in Abbott's Warringah electorate that stretches across Borogegal, Gamaragal and Gayamaygal Country, found the quality of political debate dismaying and the policy agenda years behind her first home city, Stockholm. That all changed on one fateful Wednesday when Josephson's good friend Julie said she was tired of complaining and instead was ready to take action. 'Julie said, "Seriously, let's stop talking and do something about it: are you with me?", and I said: "Sure, of course",' Josephson said.

Disengagement is unfortunately on the rise when it comes to politics in Australia, something federal MP Andrew Leigh and Nick Terrell dissected in their book *Reconnected*, with voting levels dropping and informal voting on the rise. Most of the last century saw around 19 out of 20 people on the electoral roll turn up to vote in federal elections but, despite compulsory voting, this had fallen to 18 out of 20 by 2019, and the share of informal ballots was nearly twice as high in twenty-first century elections than those held in the twentieth century. 'Democracy doesn't demand that everyone understands the nuances of every issue,' Leigh and Terrell argued, 'but democracies work best when citizens are engaged, and the level of political activity is one marker of the level of social capital in a community.' Of course I agree, having seen up-close nation-changing campaigns shift policy, funding and laws on disability, marriage equality, workers' rights and more, but I ask myself as often as you do: what could we possibly celebrate about the politics of climate change these days? All we have had is a decade or so of harmful, toxic debate that has seen off a gaggle of prime ministers, and policy progress that is, for the most part, achingly slow when compared with what is required. Take a breath before throwing this book against the wall, because I'm challenging you, and me too, to look beyond the Canberra bubbles and political argy-bargy that is too easy to moan about over dinner (or while on your morning jog). Still feeling sceptical? Read on.

Back in Warringah, Josephson wasn't alone in her dissatisfaction with how politics as usual was rolling out, but, apart from a little charity work and volunteering alongside working and parenting three children, she hadn't had much to do with it. The biggest political action she'd taken was, like most Australians, to dutifully number the boxes on the ballot papers every election day. Exploring what was around, the group of running friends investigated how to get climate action scaled up in the face of increasingly fraught national politics. 'The conclusion was that with Tony Abbott in parliament it was going to be very difficult to get anything done,' Josephson said. 'We just said, "Okay, let's see what we can do; let's grab the campaign part of it and see if we can find a candidate.' A small group of ten was built, seed funding raised, professional campaign director Anthony Reed hired, and the search was on: the group actively sought a candidate who had solid prospects of turfing the incumbent, who had the distinct advantage of, well, incumbency and all the power and influence that comes along with it. 'We were constantly communicating and meeting with community groups and they spread the word—the fact that we were professional, organised and well-funded meant we attracted high-quality candidates. In the end the candidate approached us.' From the moment former Olympian and local barrister Zali Steggall stepped into the fray, the campaign was off and running at a frenetic pace.

'Most people were laughing at us, you know, saying, "You will not do this; you can't do this"—pretty senior people everywhere would say it,' Josephson said of the campaign's early days. Not to be deterred, the group built a campaign committee chaired by businessman Rob Purves, and Josephson's time grumbling about politics was substituted with regular meetings with the candidate. Josephson enlisted her friends and family to the cause, including husband Rickard Gardell, Pacific Equity Partners' founding partner, and, with a growing network of local groups, they started taking political actions for the first time in their lives. Organising panel debates, handing out flyers, wearing Steggall's aqua-coloured campaign T-shirts all over town were all part of building the challenger's profile. As the campaign rolled on in earnest, volunteers gathered every morning at one of Sydney's biggest urban traffic pinch points, the Spit bridge connecting Manly to the CBD, holding banners, singing and dancing in support of the independent candidate. By the time election day rolled around, hundreds of people were turning up to this positive early-morning politics party. 'We'd get thumbs up and sometimes we would get boos, but generally it was really fun,' Josephson said. 'We had around 1000 volunteers on the lower North Shore: old people, retired people and young people. None of the people in our campaign were the usual political suspects, just normal people from diverse backgrounds with no agenda and no preconceived ideas about how things should be. It was really refreshing.'

The campaign remained doggedly positive, running on a strong climate and integrity-in-politics platform with $1.2 million in the kitty raised from thousands of donors large and small, and folks with serious professional skills developed the message, media and advertising strategies, creating an atmosphere of political engagement that persists in the community years after the May 2019 polling day. Zali Steggall was elected with 57 per cent of the vote, part of a new wave of independent MPs taking up a spot on the federal crossbench sparked by Cathy McGowan, former independent MP for Indi in Victoria, who ousted long-time Liberal MP Sophie Mirabella in 2013.

Josephson feels that her local community has become more energised and connected through the experience: people stop and chat more often, and are more willing to talk politics too. 'I wasn't so convinced about community work—it wasn't sort of my thing—but I've completely changed my mind about that and I'm much more engaged now in different types of community work. I'm very happy to hand out flyers and go to the climate strikes. The work that community groups do is highly engaging for people.' These days Anna's vision for community and politics is much bigger, having co-founded Zero Emissions Sydney North, a local volunteer group that is part of the national Beyond Zero Emissions network and Climate 200, an organisation supporting genuine independent candidates who commit to a science-based approach to climate action and restoring integrity to our politics. 'My firm belief is to get anything done we need to be political and get a strong crossbench, with people like Zali elected who hold to account whichever party has government. I'm spending all my spare time, money and effort on this because I think that's how we can really create change.'

ACCOUNTABILITY MATTERS

When Widjabul Wia-bal woman from the Bundjalung nation Larissa Baldwin first learned about climate change in the classroom, she quickly realised that the problem of rising temperatures and unpredictable weather patterns would have a profound impact on Aboriginal and Torres Strait Islander communities, and there were many she knew well, having visited family and friends in dozens of communities across the country while growing up. 'I remember looking at a dashboard that displayed a map of temperature variance and I thought, "This is going to mean we can't live on Country in a lot of places": knowing so many Aboriginal communities are really struggling resonated for me,' she said. 'Climate change going unchecked is going to mean forced removal of mob from their Country and so I started asking what that would mean more broadly for who we are as First Nations people in Australia, and on a global scale.'

Baldwin cut her teeth on politics from her family, her experiences building a sophisticated understanding from early on that governments have a lot of power in our lives. 'I was raised by a single mother who had to deal with being on a pension and having brothers with disabilities and that sort of stuff, so I constantly saw how much control the government had over our family in different ways,' she said. Looking from the outside, with the handy availability of hindsight, it feels inevitable that the intersections of justice, Country, climate and politics would come together for Baldwin, whom I first met in 2015 at a fellowship on community organising. Baldwin was frustrated at the lack of engagement in First Nations justice across the movement, and by 2014 had co-founded Australia's first Indigenous youth climate network, the Aboriginal and Torres Strait Islander youth climate advocacy organisation Seed Mob, with Bundjalung and South Sea Islander woman Amelia Telford. Seed Mob centres its approach to climate change on land rights, language and protection of Country, an ethos formed through more than 10,000 conversations in remote Aboriginal communities across the Northern Territory in the organisation's early days. For Baldwin, these conversations informed her thinking about how First Nations people build their political power. 'We would talk about climate change, and people would talk about managing Country and we would talk about, you know, rehabilitation, mitigation, adaptation and control. We'd talk with Elders about having a seat at the table, and one of the uncles said to me, "There is no table for us."'

Baldwin was quick to say she doesn't necessarily like or believe most politicians, something most of us can relate to from time to time. 'A lot of the time politicians only come knocking in our communities during elections,' she said. Baldwin decided to do what she could to change the story and help First Nations people get that seat at the table. She shifted to a role as First Nations justice campaigner at GetUp, and became part of a confluence of events that prompted remote Aboriginal communities in the Northern Territory to scale up their efforts to hold politicians accountable. First, the newly elected Territory government, led by Labor's Michael Gunner, backflipped on earlier pledges by lifting a moratorium on fracking in 2018, a kick in the guts to Aboriginal communities that had long been campaigning for the ban. Second, updates to the electoral roll saw people removed from the roll who didn't have an easily sourced address, a significant issue for First Nations people living in remote communities: in August 2019 Northern Territory Electoral Commissioner Iain Loganathan estimated that 16,000 Aboriginal people were missing from the electoral roll. Third, housing was one of a number of issues where the government had made big promises but hadn't delivered; communities were nonplussed. 'There wasn't a lot of accountability, even though there were a lot of Aboriginal members within that parliament,' Baldwin said. The scene was set: the 2020 Northern Territory election was going to be a cracker.

It's impossible to capture the scale of the campaign Baldwin led in the territory's 2020 poll, but it was all about maximising participation of First Nations communities, which she says can have a voter turnout as low as 20 per cent in some areas on election day. The numbers tell part of the story: 6000 Aboriginal people were enrolled to vote and 50,000 kilometres were racked up by Baldwin and crew in just ten weeks. Election day might be a couple of hours on a weekday in a remote community, so turning out people to vote was an enormous effort, people pitching in to help with translation or running around with bullhorns to spread the word that the polling booths had arrived. The results were remarkable: five of the electorates they worked in hung in the balance for two weeks after polling day. 'We reduced margins. We swung a lot of votes to Aboriginal independents, and I think more importantly, after that election, we had lots of people telling us ministers were coming to their community and meeting with them,' Baldwin said. 'Elections are so much more than just one day: you don't just choose who goes into government, but you also determine how accountable that government will be to the things and the issues that you care about.'

NEW NORMAL

Liberal politician Matt Kean has energy and environmental sectors woven into his story, both personal and political. The treasurer of New South Wales and local MP for Hornsby is a product of Sydney's northern suburbs, growing up on Darug and GuriNgai Country in a home backing onto one of the city's national parks. Kean's dad's career in the electricity sector provided well for the family, but lavish overseas holidays were off the table. Instead Kean was schooled in environmental appreciation the same way my partner and I have educated our daughters, by walking and camping in the bush. 'Our holidays consisted of walking in national parks and on weekends we were going out the back and exploring,' he said. 'Looking back on it now, I realise that it was my parents' economic circumstances which meant that we did things that were available and accessible and free.' Joining the Liberal Party at nineteen years old, Kean had the climate and energy debate as a constant backdrop to his political development, something he didn't pay a second thought to at the time. 'I never thought when I joined the Liberal Party that there would be people within the [Liberal–National] Coalition who would think standing up for the environment would be a bad thing,' he said. 'It always seemed odd that there would be people arguing against us doing the right thing for the environment.'

Soon after he was appointed to the New South Wales climate and energy ministry, in 2019 Kean travelled to Germany to make the case for the state's potential at a hydrogen conference, where the penny dropped on the scale of the opportunity we could grasp if we shift to cleaner industries, and the scale of the risk if we didn't move quickly. 'Australia is a small open-market economy and we depend on international trade for our prosperity,' he said. 'We're producing goods and materials that the rest of the world is soon going to want, so it really crystallised for me the risk for our nation in not taking action on climate change.' Kean's pitch to global investors was immediately followed by Morocco's energy minister, who not only talked up the advantage of the nation's resources and infrastructure, but the fact the government takes climate change seriously, unlike the unfortunate Oz. 'It made me realise there is a race on for investment, for opportunities and for jobs, and if we don't get ahead of the curve, we're going to miss out. I started to worry that Australia was going to become a rust-bucket state because of the lack of leadership—I'm not trying to be party political here—in general on these issues. This is the biggest economic challenge that we will see in our lifetimes, so there's a lot at stake if we don't get it right.'

Kean was highly excitable on the flight home from the conference, like a kid before Christmas cooking up ideas to get Santa to deliver the renewable-powered toys, his team making plans that could see New South Wales become the nation's first stop for clean industries like green hydrogen, green steel and green cement. He landed in a firestorm. 'I've got those images scarred in my mind of people taking shelter on our beaches,' he said from his CBD office towering above Martin Place, where large framed photographs of the state's beautiful national parks adorn the walls. 'What's great about this country is you can go there for enjoyment and family and fun. Here, people were going there to just survive. That's not the type of country I want to leave to my kids, or anyone's kids.' After walking through the streets of Sydney's CBD breathing in smoke at eleven times hazardous levels, he made a simple decision to state the bleeding obvious. 'This is not normal, and doing nothing is not a solution,' he told the smoke-infused room on the Hilton Hotel's third floor, hosting a Smart Energy Council summit. 'Longer drier periods, resulting in more drought and bushfires. If this is not a catalyst for change, then I don't know what is,' Mr Kean said. Sitting in the audience, my nose itching and eyes watering, my ears seriously questioned what I'd heard. Could this be a new moment of political possibility? After all this time?

The Mackinac Center for Public Policy in the United States uses a model to demonstrate what is within the realms of possibility when it comes to politicians enacting public policy. The Overton Window, developed in the 1990s, illustrates how politicians are limited by the ideas in society that already have widespread acceptance: if they wander outside the window they risk losing popular support. As attitudes change over time, what's possible to achieve in terms of public policy changes too. The model shows that if a policy change you want to achieve sits outside the window, it's going to be tough to get any politician who wants to be re-elected to support it. The Mackinac Center recommends starting at ground level, building support for the policy change in society and, eventually, politicians will go for it. I've applied this framework to climate change policy: in Figure 9.1 you can see where the window roughly sits. Where would you place the window right now, while you're reading this book? I hope, by the time you're digesting these pages, the window will have shifted up a row or three.

FIGURE 9.1 THE WINDOW WE NEED TO SHIFT

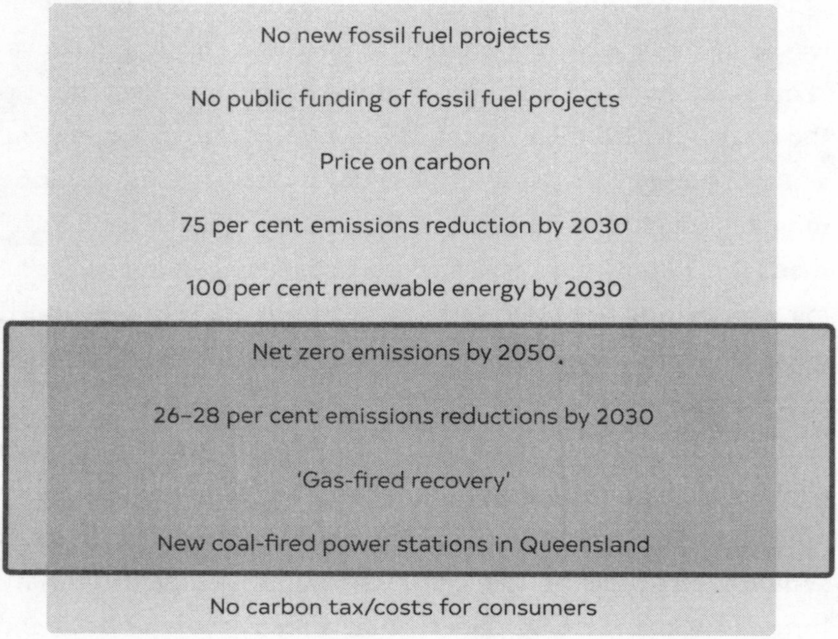

No new fossil fuel projects

No public funding of fossil fuel projects

Price on carbon

75 per cent emissions reduction by 2030

100 per cent renewable energy by 2030

Net zero emissions by 2050.

26–28 per cent emissions reductions by 2030

'Gas-fired recovery'

New coal-fired power stations in Queensland

No carbon tax/costs for consumers

Figure 9.1 demonstrates the hunch that any cynic (including myself) has that politicians will rarely take a risk, so let's wander back to the Hilton conference room on that hazy December 2019 day. It might appear that the Minister for Energy took the mother of all risks early on in his new gig, but Kean begged to differ. 'Everyone then lost their minds over that comment,' Kean said, looking back. 'But I was just reflecting what was obvious: I think everyone sort of agreed with it, but no one had really said it before.' The surge of media and stakeholder outrage was immediate, with the gnashing of teeth and fearmongering of the conservative commentariat declaring Kean public menace number one, but it was the support from everyday people that helped Kean sail through the turbulence relatively unscathed. 'We were smashed, from text messages from very high-profile right-wing commentators telling me my career was over, my views were dogshit, that I'll never go anywhere, all this kind of stuff, to my colleagues background briefing against me and whatnot. It really surprised me that the more they did that, the more support I seemed to garner from the public.' A couple of years on, Kean said his position is without question, mainstream. 'I think we've won the debate.'

Kean did a bunch in the two years he held the environment portfolio, including adding 597,000 hectares to the state's national parks and banning single-use plastics, but it's the energy plan that will see New South Wales emissions halve by 2030 in the plan's first stage, up from the previously anticipated 35 per cent. The renewable energy roadmap legislation, passed with multi-party support, will see 12 gigawatts of clean energy capacity built across five new renewable energy zones, which will accelerate closure of at least three of the states' five remaining coal-fired power stations. There's policy to build a new green hydrogen industry and government purchasing programs, and a consumer incentive scheme to stimulate electric vehicle sales too. 'If there's one thing I'm proud of, it's standing up for science,' Kean said. 'Standing up for evidence, demonstrating that the facts matter and arguing that case, which I think is the position that mainstream Australia wants their political class to be arguing.' The policies are substantial, but what's more remarkable is the politics. Legislation to support the renewable energy roadmap achieved a consensus position across the state government and opposition parties, and support was secured from key industrial unions and environmental organisations. Kean again pointed out he simply reflected community sentiment changing in his words and actions, a sign that perhaps the Overton Window is shifting in the right direction.

> We were able to legislate the biggest renewable energy package in the nation with multi-party support. You're seeing some of the strongest voices against what I've been talking about now coming around to that way of thinking—you've even got the federal government now trying to make a virtue of the fact they've committed Australia to net zero by 2050. So I think from that perspective, it's settled and it's only going to go further.

When I asked how he felt about climate change, Kean waxed lyrical on how the race to decarbonise is far larger than the industrial revolution; nations will have to recapitalise their economies to protect themselves. Again I asked how he felt, looking for any freak-out that might bubble up in the middle of the night. 'I'm very worried about the risk, but I'm excited about the opportunity that Australia can have in helping the rest of the world manage this risk.' How does he hold those conflicting emotions of excitement and worry simultaneously? 'Well, I think that I hold it because I really believe in the power of human beings to rise to the challenges that are thrown at them. Climate change is our moonshot. I'm absolutely confident we can do it, and I'm even more confident that we can lead it here from Australia.' Kean's role has been more of a political catalyst of the times than crusader, despite the media's tendency to hero-worship. 'I want people to look back on the decisions I've taken, and that we as a society have taken, and look back at that as a golden era of Australian politics,' Kean said. 'Where people did push past all the vested interests, where people did decide to put the national interest first and the world's interest first, and put in place policies that set us up for a better future: that's what success looks like.'

COMMUNITY-RUN

I first interviewed Mithra Cox back in 2004, when she was a Greens candidate in the Sydney seat of Wentworth on Gadigal Country, all bright-eyed, 24 years old, and up against a fresh field of would-be MPs including Malcolm Turnbull, who won the seat and eventually the prime ministership, only to become another political casualty of Australia's notorious climate wars. Running for the blue-ribbon seat was only one small chapter in Cox's tome of activism over two decades. She often lends her banjo and soulful voice to the cause, playing with her band, The Lurkers, at power stations and coalmine protests all over Australia. A memorable moment was performing for hundreds of people from around the world the night after they were arrested for protesting at the Copenhagen climate conference back in 2009. Middle-of-the-road climate action person she definitely is not. But these days, Cox is my local representative on Wollongong Council, after settling in the regional city with partner Martin to raise her two children. 'I can hand on heart say that, in only four years, being on council has been the single most effective thing that I have ever done in the climate movement,' Cox said. 'The amount of change that you can make when you have a seat at the decision-making table is unbelievable.'

Local government, particularly outside bustling centres, is often parodied as a small-town cousin of 'proper' politics: blinkered, backward and bumbling. But local government is where the details of our lives are governed, the things that can make or break where you love to make your home: what day the garbage is collected, whether that tree in your street should still stand, how big a new development should be, or where your faithful pooch can run off-lead. Local politics is thankfully where the political argy-bargy of the ideological national climate debate can be more easily cast aside. Elected in 2017, Cox experienced what was possible first-hand, by working closely with a couple of Labor Party councillor colleagues, carefully building support for the Wollongong area to declare a climate emergency, like Darebin Council in Melbourne had (as we heard about in Chapter 2). Declarations are helpful, Cox said, because it helps local governments step into climate action, without necessarily having a prescribed list of activities or a gigantic strategic plan in place to get the ball rolling. 'It sounds like a symbolic thing that means nothing, but what followed was really significant,' Cox said.

Soon after making the declaration, the council signed on to the Global Covenant of Mayors on Climate and Energy, which has close to 12,000 cities on board, covering around 1 billion people globally. Wollongong produces around 2,151,328 tonnes of CO_2 equivalent emissions annually (excluding 'scope 3', or indirect emissions), so by joining the Global Covenant, the council pledged to report progress each year and set targets based on the most up-to-date global agreements to cut emissions. Joining the covenant also provides councils with guidance on how to reduce emissions quickly, providing the impetus to begin shaping carbon reduction plans. A few years on and Wollongong has successfully implemented a range of climate-friendly initiatives: a food-waste collection scheme (FOGO) that saves emissions and around $3 million a year from the budget bottom line; an urban greening program that is aiming for 39 per cent tree canopy (5000 trees have already been planted); energy-hungry streetlights upgraded to carbon-friendly LEDs; new cycleways on the cards; and mining the council dump for gas to find a second use for methane that's brewing in decades of built-up city garbage. There's flipping the big switch too, with the council shifting to 100 per cent renewable electricity in 2022, cutting power bills that will then mean council can electrify everything, something Illawarra local and global inventor, entrepreneur and engineer Saul Griffith recommends. Council mowers, cars, trucks, tractors and chainsaws will go electric, and clean power will keep the lights on at sports fields, public loos and libraries.

It sounds like Wollongong is a place with a bunch of climate warriors rusted on, but this is a city with a history of coalmining and steel-making, where, as Cox said, climate change could send economic and social conditions into a death spiral, conditions that could be attractive territory for politicians to exploit when times get tough. But Cox credited the city's strong trade union history with providing a political education that pervades the appetite for agreement on council, where around 80 per cent of decisions are made by consensus. Being up close and personal also has a big impact: with only thirteen positions on the council, reasonable relationships are possible and one-to-one conversations make a huge difference to the quality of local debate. There's genuine commitment to the industrial heart of the city on all sides, Cox said, and genuine debate on issues that matter, something that is too rare at other levels of government, where the numbers are decided once only, on election day. 'I look at places like Queensland and even the Hunter, where there's this incredibly toxic binary between greenies and people working in fossil fuel industries that we've somehow miraculously managed to avoid,' Cox said. 'I think there has been a long-standing truce from all of those movements to try to find the places where we agree and to support each other on those and to navigate the complexities more quietly, rather than shouting at each other in public. And I think that has really helped to bring the community on board.' Cox added: 'What has surprised me is how much is possible, actually, and how much you can bring people with you and people that you might have thought might be diametrically opposed. If you can be polite and respectful, a bit human and nice, but also keep trying, it's astonishing how much you can achieve and how much people are prepared to come with you.'

The critical thing, Cox said, is for people to show up. When a decision to get Wollongong to set a net zero emissions target by 2030 was deferred by the council majority, Cox cried her eyes out on the cycle home, but it turned out to be a blessing in disguise. 'It came back to the next meeting and people were furious,' Cox said. 'There were 200 people at the meeting; you could hear the drums on the street outside during the meeting and they couldn't fit everyone in the room. Having all of those witnesses there meant that we were able to move on the floor . . . if they hadn't been there, we wouldn't have a 2030 target at all.'

MEETING TIME

Linda Burney is a powerful presence in person, which is why I was a bundle of excitement and nerves when I was shown into her office and offered a seat on the couch. As a proud member of the Wiradjuri nation, Burney was the first Aboriginal person elected to the New South Wales Parliament and the first Aboriginal woman to serve in the federal House of Representatives. In her long career she has never heard an Aboriginal person or community say there's no such thing as climate change, but the encouraging news is that she has seen a shift in the halls of power, showing up in private conversations and in the chamber too. 'What's really revealing is people's first speeches, their inaugural speeches: there are very few inaugural speeches that don't talk about the importance of climate change.' Reflecting on her more than 30 years' experience working on Aboriginal and Torres Strait Islander issues, she saw bipartisanship as crucial to delivering policy that's needed to address climate change, but not the type that sees little agreement and a race to the bottom as the outcome. 'It has to be a bipartisan reaching-for-the-stars effort,' she said. 'I think it's about bravery and leadership, and I think you get rewarded if you show that.'

People seem to forget that politicians aren't deities or devils, most of the time. Some are hyper-qualified, with impressive degrees and networks; some have worked in public institutions, others in the private sector; some are community, trade union or environmental champions; but they're just people when it comes down to it, as flawed as the rest of us. The only thing that sets this mixed bunch apart is power, delegated by us, to do something about the pressing issues society faces. I've met dozens of MPs from all over Australia, and while there are a few deplorable individuals for sure, the overwhelming majority of our elected representatives are decent people who share clear common values: love for the land we're on, concern for the cohesiveness and wellbeing of our society, concern for the future of our children and grandchildren. The other trait most parliamentarians share is curiosity about the people and institutions in their communities: local MPs are usually the people who can rattle off minute details about the people, industries, challenges and achievements of the communities they are responsible for. So why does our engagement with them drop off after our vote on election day?

As I mused on this disconnection point with federal Labor MP Linda Burney in her Kogarah electorate office in Sydney's south on Biddegal Country, she agreed that people often don't engage directly with their local representative. 'What astounds me is that so many people don't realise that they elect us, but they also have access to us,' Burney said. 'A lot of people would never think of making an appointment with their MP, and that's what democracy is about.' People working on the ground in community organisations are high on Burney's 'politician loves to meet' list, because they are living the issues they are speaking about. 'I tell you what, if you're a member of parliament you might get one or two people and you don't take much notice, but you get ten to twelve people saying the same thing to you, you start to notice: it's very powerful.' What was compelling for Burney is how future generations of voters are engaging. 'The first time that students were on strike, I met a group of young kids from the area; they'd just come back with their parents from Hyde Park,' Burney said. 'They were little—they weren't seventeen or eighteen, they were seven-, eight-, ten-year olds: they wanted to know what I was going to do.'

Opinion polls matter to politicians, but in-person contact really matters too. If you see a politician on the street, walk up to them, politely say hello and talk to them about your worries about climate change, and ask what they will do in your community and at the state and federal level too: these interactions make a huge difference to a politician's outlook. When I asked what people can do in their own lives to help deliver the climate action that's required, Matt Kean said the judgement calls we make should be on the values candidates stand for. 'I think that every single person can make an impact when it comes to tackling climate change or any of the other big issues of our time,' he said, 'and by acting in line with those values in the everyday decisions that they make, but particularly when they go and vote in an election. That's how we move the country.'

Sounds good to me; let's make it happen!

TOGETHER WE CAN ... *reinvigorate our democracy*

✳ Politicians are more likely to respond to the mood of the community than charge too far ahead of it. Taking action to demonstrate community expectations for faster, fairer climate solutions will be observed keenly by your local representative.

✳ Federal politics is important, but it's not the only way of achieving systemic change. Consider connecting with local councillors and state politicians as much as your federal representatives; it's where a lot of progress is being made.

✳ Politics doesn't end at the polling booth. If you want accountability and results, you need to connect with your members of parliament or council between election days, respectfully and persistently. Better still, consider standing as a candidate for election or help support a candidate who aims to deliver the change you want to see.

✳ Remember politicians are just people! You don't need to be a climate scientist or a CEO to get a meeting with a politician, and when you get an appointment, bring along a few friends who have well-developed community networks. You'll probably be surprised at how much your pollie will pay attention.

FURTHER READING

✳ Beyond Zero Emissions, 2022–, viewed 20 February 2022: https://bze.org.au

✳ Beyond Zero Emissions, *Zero Emissions: Sydney North*, 2022–, viewed 20 February 2022: https://zerosydneynorth.org

✳ Climate 200, 2022–, viewed 20 February 2022: www.climate200.com.au

✳ Climate Emergency Declaration, n.d., viewed 20 February 2022: https://climateemergencydeclaration.org

✳ Global Covenant of Mayors for Climate and Energy, n.d., viewed 20 February 2022:
www.globalcovenantofmayors.org/

✳ Griffith, Saul, *Electrify: An optimist's playbook for our clean energy future*, MIT Press, Cambridge, MA, 2021

✳ Mackinac Centre for Public Policy, 'The Overton window', n.d., viewed 20 February 2022:
www.mackinac.org/OvertonWindow

✳ McGowan, Ruth, *Get Elected: A step-by-step campaign guide to winning public office (local, state and federal)*, 2nd edn, Ruth McGowan, Victoria, 2019:
https://ruthmcgowan.com/get-elected-book

✳ Protect Country Alliance, n.d., viewed 20 February 2022:
www.protectcountrynt.org.au

✳ Seed Mob, n.d., viewed 20 February 2022:
www.seedmob.org.au

✳ Voices For Indi, *The People Are Interested in Politics: How a rural community started an Australian political movement*, Laneway Press (forthcoming, September 2022)

Part Three

WE'VE GOT THIS

CHAPTER 10

YOUR INFLUENCE MATTERS

'The call is to change the world,
and the job is entirely possible.'

—REBECCA SOLNIT

My work has involved being in thousands of meetings over the years, in person and online, from strict agendas to open, exploratory spaces, where people laugh or cry, argue and agree, commiserate and celebrate, so I can say with confidence that Stephen Moir and Carolyn Loton are masters of the art of great meeting. Partners in life and co-founders of Professionals Advocating for Climate Action, Moir and Loton are hosting one of the group's monthly online calls, where people with impressive CVs from across the corporate world, public service and not-for-profits log in to share how they are contributing to reducing our emissions through their personal and professional contributions. I've Zoomed in to check out the growing network, where people swap contact details so they can catch up one-to-one after the call: the vibe is more a networking event than community catch-up. Despite the online environment, there's no meeting fatigue to be found, which is likely why the group has grown from Moir and Loton's home-based in-person gatherings to more than 300 professionals in a couple of years. 'We seem to resonate with professionals, but everyone's welcome. You don't have to pay a joining fee, it's very open,' Loton said. 'There's a lot of people who care about the climate and the environment, who care about the problems and don't know what to do. I think what is already existing is amazing, but we feel like we resonate with people who maybe didn't feel that there was a place for them to go.'

Moir, who built a successful accounting and finance recruitment firm, and Loton, who runs her own marketing and research business, kicked off the group from their Balmain home after long-held concerns bubbled over into action in the aftermath of the Black Summer fires. 'We wanted to find ways to support ourselves to stay on track to make changes and work with other people, so we started a group of friends and contacts in the front room,' Loton said. 'When Covid hit we had to move onto Zoom and that meant the group has really grown and evolved: we've fine-tuned what we are and what we're about.' The group has four functions: self-education; sharing knowledge; making small and big changes in daily life; and influencing others in communities and networks. Every month guest speakers come along to inspire, motivate and educate the group, and every third meeting is a focused session where people share the actions they are taking, trading lessons, stories and contacts. The full professional toolkit is put to use: the network stays in touch using LinkedIn and Facebook between calls and project management software Trello is used to organise resources, actions and contacts, so it's easier for people to get started.

The philosophy is to support and encourage people to use their influence and networks to build climate-positive interventions. So far eighteen households have installed rooftop solar thanks to the recommendations from group members, who hail from five Australian states and the United Kingdom too. It's also been a brilliant way for reconnecting through a shared mission and purpose, with old schoolfriends joining up, bringing their friends along, spreading the encouragement around. A big achievement the couple noted is when the online network spilled over into real life, reaching out through their networks to get 190-odd people to a business breakfast held by the Nature Conservation Council of New South Wales. 'It's nice to feel we're doing it together,' Loton said. 'I get nervous every month because it's getting so big, and I find it a bit daunting sometimes, but that face-to-face event was a real high.' When I asked what the potential for influence was through the group, both could see possibility in spades. 'We've got a wide range of roles in the group; it's everything from finance people to doctors to university academics, so people have got the potential to influence,' Loton said. 'I think the group gives them a strength to then go back, particularly to them trying to influence in their workplaces, because a lot of group members are reasonably senior in their professions.'

NO KARDASHIANS NECESSARY

Sociologists have spent decades trying to understand how ideas and action spread, with a vast field of social network theory and analysis likening the spread of ideas to a virus. For a communicator or campaigner, seeing something 'go viral' is so, so satisfying, because it means your message is taking on a life of its own. Yet when it comes to creating the enormous suite of changes we need to decarbonise in a short timeframe, spreading information is only the beginning: we need long-held beliefs and ingrained behaviours to shift too. Damon Centola, a sociologist who studies the science of social networks of both the online and face-to-face kind, challenged nearly a century of understanding of how social networks cause behavioural change to spread in his fascinating book *Change: How to make big things happen*. Centola distinguished between two types of spread: simple and complex contagions. Spread of information is a 'simple contagion': ideas spread like a virus but lack long-term effect. Social change is quite a different beast, Centola wrote, because simple exposure isn't enough to 'infect' and spread it around: an individual has to make a decision to accept or reject a new behaviour or belief. As these types of decisions are often complex and emotional, spreading them around requires a deeper process: they are therefore 'complex contagions'.

When it comes to influence, we've been taught by the media that celebrities and elite members of the business community or politics have their hands on the reins, steering the way we think and act. Marketing departments across all sectors have influencer strategies, hungry for the prized celebrity endorsement or retweet. But Centola's investigations over a decade have found that while influencers are useful for spreading the simple contagions like information or gossip, complex contagions need specific conditions to spread. 'When it comes to social change, the myth of the influencer obscures the real pathways that have led challenging and even controversial social, commercial and political initiatives to succeed,' Centola wrote. Centola's research stretches from the adoption of new health-management tools to analysing the inexplicable explosion in the uptake of Twitter in 2009, and the spread of the #MeToo and #BlackLivesMatter movements, and his conclusion is that influencers are more likely to endorse change, be it a new product, a candidate for election or a cause, *after* a change has reached its exponential adoption rate, not before.

Centola concluded that highly connected influencers are like a hub with spokes radiating out, and they appeal to our ideas of the hero leading us to a win for all. But it is the interlocking ties that permeate the *peripheries* of networks that make the biggest difference to spreading behavioural change at scale. The places we're looking for are on the edges of networks rather than at the centres, where social ties come together to strengthen bonds between groups, whether that be between families, partnerships across organisations or solidarity within nations. 'Because the network periphery is so large and unexceptional, it can appear less significant than the networks of highly connected social stars,' he wrote. 'But the truth is just the opposite: when it comes to social change, the network periphery is where all the action is.' I hope you're feeling more at the centre of the action, even if you're not the meme of the day on TikTok.

Centola wrote that wide bridges—multiple overlapping connections between social groups—are essential to build the necessary trust, credibility and legitimacy an innovation requires to take off. These wide bridges provide more opportunities for people in a group to observe things from more people; they increase trust and reduce risk for people who are considering adopting the change. You've probably noticed this phenomenon in your workplace or community: the same people seem to pop up in multiple groups. According to Centola's research, this is a strength, not the duplication of efforts you might assume. 'Wide bridges are not about reach but *redundancy*,' he wrote. 'They allow people on both sides of the bridge to hear the opinions and recommendations of multiple peers and colleagues, and to discuss and debate ideas with them. Wide bridges mean stronger ties.'

Climate- and sustainability-focused action initiatives of all kinds are springing up all over Australia, whether they be profession-ally focused, like Work For Climate, Climate Salad's growing network of entrepreneurs, or place-based organisations like Green Connect. These initiatives stand to be more successful if they are building wider bridges, something Professionals Advocating for Climate Action is doing through its monthly meetings, engaging with existing networks rather than spending energy trying to enlist movie stars.

VEGEMITE MAN

With ten years of community campaigning with Friends of the Earth in Melbourne under his belt, Leigh Ewbank is well placed to discuss influence in communities. 'I often say I'm in the busi-ness of building community,' Ewbank said. 'A lot of the work involves getting out into the regions, particularly communi-ties that might not have ever met a climate campaigner, and that's one of the most exciting parts of the gig.' Friends of the Earth is a global movement, the Melbourne part of the network emerging from a community campaign in the early 1970s to stop a nuclear power plant from being constructed on French Island, Victoria's largest island, in Western Port Bay. Community-driven campaigns are part of the group's DNA, said Ewbank, who is the coordinator of Act on Climate, a campaign collective, which is just another name for a bunch of community members who share a passion for the climate crisis and doing something about it. In 2021, Friends of the Earth released the People's Climate Strategy, developed as the global pandemic gripped Victorians in wave after wave of lockdowns. Volunteer coordinator Anna

Langford and Ewbank brought people together from Portland in the state's south-west to Mallacoota on the east coast, from the heart of Melbourne to the Murray River border, in forums, on social media and using surveys, everyone having their say on the next chapter of Victoria's Climate Act. Five years of tireless advocacy by Friends of the Earth with unions and community groups won big: in March 2022 Victoria's State Government announced a colossal offshore wind target, which promises to create up to 6100 jobs in fifteen years. About half of these jobs will continue after construction finishes.

'One of the things I love about the work is that because it's so resource-constrained, at the grassroots end of the spectrum, you have to be very creative,' Ewbank said. 'Those tough conditions are part of the recipe for coming up with the coolest, zaniest strategy and tactics that will have an impact.' Building influence has involved knocking on doors, holding town-hall meetings, writing articles and following politicians around, popping up unannounced at speeches and events (it's called 'bird dogging'). One memorable intervention went back in time to engage in a public inquiry set up by anti–wind power members of federal parliament. In an exercise of pointed political satire combined with some cheeky street theatre, the 'Flat Earth Society' congratulated the pollies on their anti-wind witch-hunt by making formal written submissions to the inquiry and dressing up in mediaeval garb, howling down politicians who were wallowing in the past. But it's the community work that has seen big outcomes. Between 2010 and 2016, Friends of the Earth (the acronym is, hilariously, FoE), engaged with 73 towns across Victoria where there were licences for coal-seam gas and unconventional gas exploration and extraction. Each community formed local action groups of people who knocked on doors, asking their neighbours if they wanted their area to be gas-field free. In each of those communities, there was 96 to 97 per cent agreement, influencing a permanent ban on coal-seam gas and fracking in Victoria made law in 2021. 'They declared themselves gas-field free, not waiting for the government to do it,' Ewbank said. 'There was so much power built in this community campaign that when the Daniel Andrews government in Victoria put forward the bill, the Nationals voted for it, Labor voted for it, the Liberals voted for it, the Greens voted for it—everyone voted for it. This was a clear demonstration of the community's will and a demand for what the government should do, and they delivered.'

For Ewbank, influence is as individual as personal preference, and your unique strengths can be harnessed however they manifest. 'I'm an absolute introvert, but climate change scares me so much and I want to do something about it, so it forces me to break out of that when I'm in campaign mode,' Ewbank said. 'I can emcee a town-hall meeting or do a presentation and get people psyched, but I would much prefer being, you know, with an individual having a conversation—for me, that's really comfortable.' Ewbank reckoned the key to building influence is listening, not proselytising. 'I never, ever push a barrow when I'm in a community; it's about having presence and demonstrating that people that are committed to social justice and climate action are regular people,' he said. 'We can have a beer at the pub or a pie at the bakery and just get along on a human level, and even though we might have disagreements about technologies, whether it's wind farms or solar, or even values, we still can connect on that human level and find common ground. I think in the age of social media, where we're seeing algorithms intensify social bubbles, there is a really crucial need for us to kind of challenge that tribalism and remind people there's a lot we have in common.' Sometimes it takes an Aussie icon to help remind us, so bring on the Vegemite.

Ewbank loves Vegemite, like just about every Australian: about 1.2 billion serves are spread every year and around 80 per cent of our nation's pantries have a jar on the shelf. 'I can't go past a bloody Bakers Delight without getting a Vegemite-and-cheese scroll, I just love it,' he said. (For me, it's got to be on toast: white bread with loads of butter and just a scraping of the stuff is a cure-all.) But how on earth could Vegemite help with climate solutions? This story is a tale of opportunity and accident, perhaps, but it also shows how connection and community can make its mark in the times we're in. When Hydro Tasmania proposed a 200-turbine wind farm in the Bass Strait off the coast of King Island in 2013, it was the biggest proposed in Australia at the time. The proposal landed in the small community like a lead balloon, something the anti–wind power lobby was keen to exploit. Ewbank, who had cut his teeth coordinating communities keen to see more climate-friendly wind projects in their towns, was contacted by a King Island local wanting to work out how to combat misinformation on the impacts of the new technology. 'It's a small island. It's not very populated, and King Islanders are very proud and passionate about their community too, so you can imagine how this could become quite a heated issue,' Ewbank said. When community information sessions were announced, Ewbank scraped together the cash and flew 250 kilometres south to visit the community over a couple of week-long stints. A friend reminded him it was going to be chilly when he arrived, so she passed him a thick, black woollen hand-knitted jumper with the Vegemite logo splayed larger than life across the torso. 'I was like, "Oh bloody hell, why not?"' Ewbank said. 'It's going to be obvious to everyone that I'm not a local so I may as well just be really upfront about it.'

Ewbank, jumper in his backpack, stepped off the plane and immediately got cracking, visiting the local library where he learned about the rich history of King Island, including how many members of the small community fought in World War II. Ewbank wandered around town making connections with the locals at the cafe and the bakery and popped into information sessions on the development put on by Hydro Tasmania, quietly tapping away at his laptop, observing the people and the process. 'The second time I got back to the island was one of the first times that the environment or climate movement had had someone going head-to-head with the anti–wind farm lobby at the same time in the same place,' Ewbank recalled. A leading anti-wind campaigner was in town when he arrived, who was top billing at a No TasWind Farm community forum, where the misinformation was likely to be laid on pretty thick. 'It was just like the propaganda show for wind-turbine syndrome: they're loud and they're going to cause all these problems—people will be getting sick, the usual stuff we've come to expect from the anti–wind farm crew,' Ewbank said. 'So I decided to go to the meeting, and just do what I've been doing the whole time, just document thoroughly what was discussed and be there and see what people were saying.' Rocking up to the meeting a little late thanks to a tasty pub dinner, he had barely sat down before he was ejected from the meeting. 'This guy did the kind of standover man thing and towered over me, while I was getting my computer ready, and he just said, "You're not welcome here. You've got to get out; you've got to leave," and marched me out of the town hall.'

FIGURE 10.1 WHO IS THE 'VEGEMITE MAN'?

Leigh Ewbank, the "Vegemite man" expelled from the No Tas Wind Farm Group's meeting.
"The only reasonable threat I posed the meeting
was an intellectual one," concluded Ewbank.

Source: Courtesy of Leigh Ewbank and the *King Island Courier*

It was a tense moment for sure, but a recording of the event revealed that Ewbank's efforts building connections and trust in the town had been crucial, with the move to push him out of the meeting causing a stir. 'People in the audience started saying, "What are you doing, kicking out the Vegemite man? He's all right,"' Ewbank said. That was the night before Anzac Day, so on the morning of 25 April, Ewbank was up early, attending the Dawn Service at the local cenotaph before heading to the RSL, where word had travelled fast. 'People started coming up to me and asking if I was the Vegemite man, and saying, "I just can't believe they kicked you out of that meeting last night!" People who weren't even at the meeting heard about what happened.' The story was on page three of the next edition of the *King Island Courier*. After only a couple of weeks, an outsider greenie from Melbourne had become just another good guy anyone could have a yarn with: the Vegemite man had arrived. 'I think it really showed the true face of the anti-wind organisers,' Ewbank says. 'They weren't open-minded, they weren't even prepared to have someone there who could challenge their views, or even just sit there quietly and document them. They were in the business of proselytising, but we were in the business of listening. It did turn the tide a bit in the community.'

Ultimately the development didn't proceed in its proposed form, but these days the community hosts the King Island Renewable Energy Integration Project (KIREIP), a hybrid off-grid power system using a combination of wind and solar, replacing 65 per cent of the diesel that was used for electricity generation, a demonstration project that is the first of its kind in the world. As for the woolly jumper, it was Ewbank's constant companion over the next couple of years, helping break the ice in communities across Victoria, before it was misplaced on his travels drumming up engagement for climate action and renewable energy. He'd lost his icebreaker but not his superpower. 'A lot of the work does involve me going into communities where I am a blow-in, so I may as well do it in style and be proud and obvious about it,' he said of his Vegemite days. 'It's a very popular condiment, so it tapped into Australia's larrikin spirit: I think it helped me get in and connect with people.'

We could spend hours crafting our own superhero suit, but the best I have ever managed to knit is a scarf (no stripes), so perhaps our time is better spent looking for ways we can connect and influence.

TALK IT OUT

If you've ever thought you're not a good speaker or writer, you're among friends: this is the top barrier to action identified by the 24 per cent of Australians who are most alarmed about climate change, according to the Climate Compass research outlined in Chapter 1. Dr Rebecca Huntley, who literally wrote the book *How to Talk About Climate Change in a Way That Makes a Difference*, made the call that this is one of the hardest topics to talk about, more difficult than sex and politics, more on the level of death and depression. 'The silence around climate change in our daily lives works to decrease our sense that it is a real and imminent threat, which in turn affects everything from our voting decisions to our purchasing choices and general behaviour,' Huntley wrote. 'This goes for the silence in our daily lives as well as the lack of consistent attention to it in broader conversations at the community and company level and in the mass media.' I'm an extrovert, which means conversations with almost anyone on anything are pretty easy for me, but when it comes to climate change it often feels like my wings are clipped. When I've quipped about climate change in my gym class or expressed my frustration at a family gathering, the silence hovers awkwardly, disrupting the flow. We must break the climate silence, Huntley argued, to help circumvent the psychological barriers to action that we are facing. If we're in the business of creating systemic change, we need to tackle this tricky topic in conversations with our nearest and dearest and people in networks we have the ability to influence. So how do we go about it? What does it take to have a productive conversation on climate change? Is there some secret sauce that has the power to transform friends, family members or work colleagues?

Seeking advice, I turned to the leaders of Climate for Change, a unique organisation with expertise in conversations that transform rather than terrify. Co-directors Lena Herrera Piekarski and Jane Stabb describe the fundamentals of how the process works: group conversations held in small groups in people's homes, more Tupperware party than committee meeting. The images flash through my head of afternoon tea parties I was carted around to when I was small, full of 1970s fashion and durable plastic on display in every colour to match the decor, the conversation flowing as the host racked up the orders. Climate for Change uses a similar format, but the content covers far more challenging terrain. 'We know what changes people: it's not hard science, but emotion and connecting with someone that you trust,' Herrera Piekarski said. Here's how it works: you invite a small circle of friends over (or to an online call), where you are shown a fifteen-minute video on the problem by a facilitator, then you are guided through a conversation that unpacks what the world could be like in 20, 30, 50 years' time, delves into the pathways we might choose for our communities' future and, crucially, how we feel about it. 'It's drawing a connection between the scientific projections and what that means for us, as humans, in our families, our friendship circles, in our networks,' Stabb said.

> When I first came to this organisation, there was some magic about it that I couldn't clock and there's still something that's mystical about it, because if you talk to someone who's been through one of these conversations, these are life-changing, transformative, conversations. We know that roughly 70 to 80 per cent of people across the voting population want to see more action from their political leaders. That's a thing to know, intellectually, but it's a different thing to feel.

Crucial to helping the conversation along is the trained facilitator, who is usually from within the host's network or a step or two removed. It's designed so people coming along feel the facilitator is more like them, with common interests, experience or social ties. It's intuitively rolling out one of Centola's principles of *relevance*: when behaviour change requires feelings of solidarity and loyalty, similar sources of reinforcement will help inspire action. There are more than 10,000 people involved in Climate for Change, including Suzanne Mungall from Kingaroy, whom you met in Chapter 1. Mungall found Climate for Change after searching online for climate action in her region, and these days she's an active volunteer, driving hundreds of kilometres to facilitate group conversations, as well as meeting up with people one-on-one to talk. 'One of the important things that the climate movement provides is a community of people who are going through that transformative emotional journey, that connects them and provides a space for them to make meaningful change,' Stabb said. 'Whether that's calling their MP or switching their super, they sign on to take action in a way that's meaningful to them.'

Herrera Piekarski and Stabb recommended a few key principles for having a productive conversation, which you can see in detail in Figure 10.2 and the box 'A note on questions'. But the shorthand version is pretty simple. First, start with a personal connection: establishing rapport by being generous with what you do share and agreeing on areas of common ground, particularly on values, is important to building trust and mutual respect. Second, listen and be present: the rule we learn as organisers is listen for around 70 per cent of the time or more and talk for 30 per cent or less. The more you bombard with facts and statistics or express your own frustration, the less the person you're talking to will feel seen or heard. Third, you don't need to be an expert; in fact, long lists of

FIGURE 10.2 A GUIDE TO CONVERSATIONS ABOUT CLIMATE

START WITH AN OPEN QUESTION

'How do you feel about climate change?'

Remember to listen more than you speak, without judgement:
'Tell me more about . . .'

Validate their opinions. Find common ground. Tell your story to show why you want action. Paint the picture of success.

Find out what holds them back

COMMON BARRIERS

There are a few common barriers to taking action that consistently came up for different groups of people. Here are some ways to overcome them.

'I don't know enough'

It could be helpful to dig a bit deeper to find out more detail on what they need help with. *'What do you feel uncertain about?'*

Share resources you trust with them. Acknowledge that it can be confusing.

'I don't know what would make the biggest impact'

'I feel I do a lot for the environment in other ways'

Help them understand the most meaningful action they can take: *'To prevent climate change getting much worse, it's going to take a huge transformation. That means we should put our efforts into calling for government and businesses to change the system, so our individual and consumer actions are part of a broader push for change that takes us in the right direction.'*

'I'm not a good speaker/writer'

Reassure them that they don't need to be an expert—just a person who cares deeply about our future. *'I was nervous to write/call/speak too, so I reached out to a climate action group for help, and asked some friends to join me. We all really enjoyed it, and it felt less of a burden when we shared the load.'*

'Australia needs the coal/oil/gas industries'

'Did you know that more than 30 per cent of Australia's electricity already comes from renewables? We're in the sunniest country in the world, and one of the windiest. If we had policies that made the most of this natural advantage, our economy could be so strong!'

Source: Adapted from Climate for Change, www.climateforchange.org.au

ENCOURAGE ACTION!

Ask them: are they ready to take action?

Affirm their concern, and your shared purpose/perspective:
'I see you care about future generations; like you, I'm worried my children will find things difficult.'

Ask for a commitment.

What is the action?

Can you take action together?

FACING OBJECTIONS

Conversations to create change will often result in people questioning or arguing with what you say.

A useful framework for handling these objections is:

Explore: what is behind the objection?
(Is it a lack of knowledge, misinformation, etc.)

Empathise: acknowledge their position, even if you don't agree—*'I understand . . .'*

Elevate: lift them up with facts or a story of hope—*'I thought that too, until I learned . . .'* or *'Did you know . . . ?'*

facts and statistics are more likely to push people away, so talk about how you feel instead. Fourth, have realistic expectations: it's reasonably unlikely that one conversation will see a mildly concerned friend or co-worker chuck in their job for full-time Extinction Rebellion protesting. But the more you engage and establish that you're a person who cares about them and the more support you show for the journey they are on, the more chance they'll stay the course. 'I often start conversations about climate change talking about something completely different,' Herrera Piekarski said. 'It might be about the experience of the summer or something of value that we share or are inspired by. If I can find that one thing that they care about, it might get them to shift in the conversation, despite them maybe not being on the climate change bandwagon yet.'

What about people who object, or want to argue with you about climate change? The fear of triggering conflict is one of the biggest hurdles we need to overcome to get started, so who better to ask than Mithra Cox, the Wollongong councillor who has had thousands of conversations with locals on the street, where some will be angry, others will come on the offensive. 'In those situations, I bring it back to something that's really personal that people can't attack you on, which is my feelings,' Cox said. 'Rather than reacting to whatever it is they are saying, I say that I am really worried about climate change, and it's something that keeps me up at night and fills me with terror: no one can argue with how I feel. People soften when you give them a little bit of your vulnerability; it's a natural human reaction to be a bit kind, and people will often wind back and try to agree with you.' When it comes to outright climate-deniers, Herrera Piekarski recommended disengaging. 'If that's your own family member or your close friends, that can be really hard, because you want them to see the world the way you see it,' she said. 'But at the end of the day, if it's a relationship that you want to keep, the relationship is more important.' Your time and energy is far better spent, she said, by having quality chats with people who are concerned, but not yet active.

A NOTE ON QUESTIONS

Asking questions demonstrates you're engaged and present in a conversation, and is the best way to build rapport and trust over sticky issues like climate change. In **The Art of Powerful Questions**, Eric E. Vogt, Juanita Brown and David Isaacs put a compelling case for considering more carefully the questions we ask in our interactions, providing insights for climate conversations we're all going to have more of. 'Questions open the door to dialogue and discovery,' the trio wrote. 'They are an invitation to creativity and breakthrough thinking. Questions can lead to movement and action on key issues; by generating creative insights, they can ignite change.'

But not any question will do: to ensure our questions are powerful, we need to consider the assumptions we are embedding in our queries. Are we loading up our questions with our worldview, trying to cleverly railroad? Or are we genuinely trying to find out where someone is at?

Less powerful questions—the 'yes/no' type—narrow the scope of possible answers and can close down conversations, so avoid asking something that demands agreement (or disagreement) when talking climate with someone you're hoping to influence. 'Who/where/when/what' questions are useful for drawing out information, but these are easily surpassed by the more powerful questions: 'how' or 'why'. Asking questions using these words opens minds, allowing for more reflective thinking and deeper conversations.

TOGETHER WE CAN... *be influential*

✳ The networks you have right now are powerful, so consider how you can seed your networks with ideas for climate action that have relevance for people.

✳ You don't have to be a celebrity to make your mark. Building wide bridges of people across different networks helps build trust, legitimacy and credibility for new beliefs and behaviours to be introduced.

✳ It's about listening and connecting, not proselytising. Influence will come when the person you're in conversation with feels seen, heard and valued for who they are and how they feel.

✳ We need to talk about climate if we are going to achieve the enormous scale of change that is required. Overcome the awkward by connecting on values, talking about your emotions and keeping it real. Remember: you don't need to be an expert in carbon sinks, electrons or rewilding—be your authentic self and build trust.

FURTHER READING

✳ Centola, Damon, *Change: How to make big things happen*, Little, Brown Spark, New York, NY, 2021

✳ Climate for Change & Climate Compass, 'Climate conversation guide', n.d., *Let's Create a Climate for Change*, viewed 21 February 2022:
www.climateforchange.org.au/download_guide

✳ Friends of the Earth: Melbourne, 'Act on climate: Ensuring Victoria becomes a leader on climate justice', n.d., viewed 21 February 2022:
www.melbournefoe.org.au/climate

✳ *Professionals Advocating for Climate Action*, 2021–, viewed
21 February 2022:
www.professionalsforclimate.com.au

✳ Vogt, Eric E., Brown, Juanita & Isaacs, David, *The Art of Powerful Questions: Catalyzing insight, innovation and action*, Whole Systems Associates, Mill Valley, CA, 2003:
www.scribd.com/document/18675626/
Art-of-Powerful-Questions#download

CHAPTER 11

YOUR UNIQUE STRENGTHS

'Who better to know what to do at this time,
in this place, with your knowledge, than you?'

—PAUL HAWKEN

What would you get if you pulled a ruck of AFL players, a corps of retired diplomats and a renegade Tesla-driving Extinction Rebellion activist into a fight for the future of our planet? It would be a pretty wild team, one that's always growing and changing, where everyone has a role to play. The amazing thing is that this is precisely what's happening, right under our noses. In Climate Action Starts Here, at the end of this book, and on my website I've had a go at building a list of organisations pushing for rapid emissions reduction, no new fossil fuel projects, faster renewable energy uptake, more support for communities to prepare for, and recover from, extreme weather events, while ensuring justice and equity are delivered through the transition that's coming our way. There are so many ways to be part of efforts to ensure systemic solutions for the climate challenge that it's becoming easier than ever to align your experience, knowledge, skills and passions with pushing for climate action across our economy, our society and our culture.

Take Nicola Barr, a rising AFLW star, who joined the Western Sydney Giants as the number-one pick in the league's inaugural draft. Barr was awarded the 2016 Mostyn Medallist as the Best and Fairest player in the Sydney competition and nominated for the NAB AFLW Rising Star Award in her first season in 2017. Nicola and I caught up while she was in isolation, recovering from a thankfully manageable bout of the dreaded Covid while trying to keep up with a disciplined training regime in preparation for the upcoming season. We discussed the impact of a changing climate on sport and the potential for Aussie Rules to help accelerate solutions. Given AFL, its origins said to be in the Aboriginal game Marngrook, is more of a religion than a simple weekend pastime for millions of Australians, there was plenty of ground to cover in our conversation. Barr always played sports as a child; her father's job in construction saw the family move around to various far-flung locations like Dubai and Singapore before eventually settling in Manly, on Sydney's north shore, when Barr hit her teen years. Sport was the great unifier and social conduit, the bridge to connection. 'Having moved around a lot as a kid, sport was always my way in to meet new people; it breaks down a lot of barriers,' Barr said. 'I always looked to sport to make friends and feel a sense of belonging.'

Climate change was always something that bothered Barr, whose family's adventure-style holidays—biking, hiking and running around such far-flung places as Germany, Patagonia and Jordan—cultivated a big love of the natural world. 'I realised how much I loved being in nature, being outdoors and experiencing things, by moving my own body,' she said. 'I absolutely love doing that, it gives me energy.' A growing awareness was compounded by the visceral experience of training at Sydney Olympic Park in the city's west through the Black Summer fires, the heat and smoke overwhelming, and seeing games cancelled because of severe storms. 'There were all of these events and I started looking at everything that I enjoy doing, and I thought this isn't going to be available to people in the future if things keep happening this way,' Barr said. 'I started reading a lot more, listening to a lot of podcasts and engaging in conversations with friends who also care about it as well, then I started looking for a way to be involved in making a difference.' It was perfect timing. Barr's search for ways to ramp up her climate action quickly led her to St Kilda's Tom Campbell and retired North Melbourne and Port Adelaide player Jasper Pittard, who were forming up a brand-new team: AFL Players for Climate Action, which launched in 2021 with more than 260 professional footy players signing on.

'Other than being an early adopter, bringing my own shopping bags and having a KeepCup, I didn't feel like I was doing enough; I carried a bit of climate guilt around all the time,' Campbell said. 'I didn't really know how to do it, which I think is a pretty common experience if you're feeling paralysed by just how massive this issue is.' Campbell and Pittard talked climate in the locker room when they both played for North Melbourne, but assumed that, because their fellow footy players weren't doing the same, they weren't interested or terribly concerned about the problem. That all changed during the Black Summer fires, with the disaster sparking conversations across the club, giving the pair a big shove to do more, beginning to build the group a few months later. When Campbell and I chatted, they'd already surveyed 600 professional-level players across the sport, finding 92 per cent are concerned about climate change, with varied reasons for why. 'We've got players who are new parents, who are really concerned for the future for their kids,' Campbell said. 'We've got players who have a deep connection to Country or place and they're really concerned that will be lost in some way with accelerating climate change. And we've got people who simply just care about the future of our game, where people might not be able to play it in the country town where they come from in the near future.'

The fresh AFL climate action team's inaugural members includes league greats like 2021 AFL premiership player Ben Brown, stars Dyson Heppell, Jordan Roughead and Luke Parker, and AFLW luminaries including Daisy Pearce, Erin Phillips, Darcy Vescio and Nicola Barr, who have all pledged to tackle climate change in their personal lives, at clubs and across the industry. Two initiatives that kicked off the team are partnering with GoNeutral, a start-up that offsets transport emissions, which are significant in a national sport that plays across a big country every year: the average transport emissions per player for the AFL men's 2021 season were around 8 tonnes of CO_2 equivalent, and for AFLW players it was 3.7 tonnes, a figure that will increase as more teams are added and travel demands grow in the women's game. More than 180 players from the group signed on to The Cool Down, an initiative sparked by former Rugby Union Wallabies captain and avid conservationist David Pocock, which brings together professional sporting men and women supporting cuts to greenhouse gas emissions at least in half by 2030 and to reach net zero emissions before 2050, in line with scientific demands. Australian cricket captain Pat Cummins signed on to The Cool Down's open letter, along with tennis great Mark Philippoussis, swimmer Libby Trickett, netball star Liz Ellis and Formula One champion Mark Webber: the list of big-name stars is impressive, and there's also a host of talented sportspeople from softball, taekwondo, trail running, trampolining and more. In February 2022, Cummins stepped things up a notch, launching a new group, Cricket for Climate, that is starting off with a campaign to see all cricket clubs install solar panels.

For Barr, AFL Players for Climate Action has quickly become a place where she can learn, connect and combine her considerable sporting talent with her passion for fixing the planet. 'I'm not a massive player: I don't have a huge social media following—nothing like that,' Barr said. 'I just realised, why wouldn't I start talking about this more vocally, why wouldn't I do this?' Barr is taking steps to encourage her club to reduce its impact, talking more with teammates, friends, colleagues and family too, and recently stepped up to join the group's inaugural governing board. 'I love my football, winning and being an athlete, but now I also feel connected to a bigger issue, and I feel like I can make a difference with this group. It's given what I do every day a lot more meaning,' she said. 'I really wanted to know how I could contribute to this space and make a difference: the AFL has such a huge impact across Australia, so if we can continue to use that platform and make a bigger difference, that's really exciting. Seeing Tom and Jasper's passion for what they want to do, as well, has pushed me to do more.'

As far as Campbell's concerned, teamwork is where the magic is, on the playing field and for the planet too. 'The role of sport is to normalise climate action with footy fans and communities right around the country, and that's what we aim to do with this group. We've managed to connect with players from both competitions all around the country who are passionate for different reasons and have different strengths and different levels of involvement,' Campbell said.

I've spent my whole life being a part of teams; it's where I feel most comfortable. In terms of action on climate change, I was an individual and I didn't know where to go or what to do. I think once I joined the team—and that's not just the AFL Players for Climate Action team, that's this big network of people and communities that are trying to work on the solutions that are already at hand—things become easier. My advice to anyone would be to join a team and become a part of working together to solve this problem.

DIPLOMACY CALLS

As the footy stars were lacing up their climate action boots, former Australian ambassadors, consuls-general, trade commissioners and aid workers were busy harnessing skills built over decades of diplomacy to advocate for urgent action to protect the globe. Richard Mathews served in exotic locations including the tiny nation of Brunei and Greece's capital, Athens; established a post in Makassar, capital of South Sulawesi, eastern Indonesia; and led sections focusing on India, Sri Lanka and Bangladesh: all par for the course in a 30-year career with the federal Department of Foreign Affairs and Trade. 'You might get asked to help somebody who's trying to sell widgets, or you might have to go and look for a lost yachtie who's disappeared off the radar in some islands in eastern Indonesia,' Mathews said of international postings. 'The job is basically to represent Australia in anything and everything you could imagine.' From hosting the prime minister and senior ministers to liaising with the US's Central Intelligence Agency and Israel's Mossad on security during the Athens Olympics, Mathews said strengths useful for the job include an appetite for policy, solid general knowledge, bucketloads of patience, critical thinking and, most importantly, excellent people skills. 'Most diplomats are fairly serious types, but they're not risk averse or change averse, because, in our career, you have to be very resilient and open to change,' Mathews said. As he talked through the strengths unique to a diplomat's working life, it was like hearing a run-down of a climate advocate's skill wishlist.

Mathews was a long-time supporter of environmental action, joining Friends of the Earth for a time in his university days and regularly donating to environmental charities, but when his first granddaughter was born in 2019, his concern shifted gears. 'I started thinking about the future a bit more, and when I started reading about climate issues I thought, "Oh my god, what sort of world are we leaving behind us?"' he said. 'I was thinking about my granddaughter's future and asking, in 80 years' time when she's an old woman will this planet still be habitable? That sort of thinking led me to decide to be more active.' Retiring from the public service in early 2021, Mathews had a new-found freedom to work on his passion for climate; searching for ways to lend a hand, he joined his local Australian Conservation Foundation action group. It was only a couple of months before he began redeploying those handy skills honed by decades of diplomacy, building a collaboration with scores of former colleagues. Together they began speaking up for sensible foreign policy and climate action, an area where Australia's government had unfortunately flipped from leader to laggard on the world stage.

Mathews found himself engaging typical diplomacy skills in communications and negotiation to develop an open letter to the federal cabinet (see the box 'An open letter to the Australian Prime Minister and members of the federal Cabinet'), followed by the development of a full-scale climate-focused foreign policy for Australia. Under the banner of Australian Diplomats for Climate Action Now, more than 90 former diplomats—including 38 former ambassadors, plus an array of concerned citizens with impressive titles like high commissioner, consul-general, senior trade commissioner, humanitarian aid worker and department head on their CVs—contributed to the policy framework, which spells out how Australia should meet its international obligations. 'We want Australia to be out there as a world leader in climate policy, showing the rest of the world how to do it and helping the rest of the world, our region and our neighbours manage this transition,' Mathews said. Enshrining the COP26 commitments from Glasgow in legislation, rapidly reducing our CO_2 emissions, doubling Australia's allocation of climate finance, recommitting to the United Nations Green Climate Fund and supporting developing nations, especially our neighbours in the Pacific, as sea levels rise are high on the group's agenda. 'There's a lot of other considerations for small, low-lying states: morally and politically Australia has a lot of responsibilities in that area.'

Mathews said he's been impressed by the sheer level of concern among his former colleagues, which drives the collaboration's energy. 'There would be a large proportion of our members who are dyed-in-the-wool conservative voters, and there are some hardcore lefties in our group too,' he said. 'We've all come together because of this shared concern about the future.' Mathews said he is certainly worried about climate change but is also optimistic that Australia can act on the solutions we need. 'What our group does is just a drop in the ocean, but I've realised there are groups across our society, across all of our communities, that are concerned and are taking action. It's just the most amazing thing; I never realised how active Australians are.'

AN OPEN LETTER TO THE AUSTRALIAN PRIME MINISTER AND MEMBERS OF THE FEDERAL CABINET

As former diplomats we are deeply concerned that Australia's key strategic and economic interests are at risk because of our failure to date to commit to a target of net zero emissions by 2050. This lack of commitment is particularly concerning to those regional partners for whom climate change already poses a clear existential threat. The United States and other key partners in Europe and around the globe are increasingly voicing concerns that Australia is not pulling its weight on climate action. Australia's inertia on commitments undermines our credibility as a regional partner; it undermines our reliability in the minds of our strategic allies; and it will cost us dearly as trading partners seek to impose carbon tariffs on imports of our goods and services. We fear this inertia will undermine many of the strong international relationships we have built up over decades.

Source: Excerpt from the open letter sent by Diplomats for Climate Action Now to Prime Minister Scott Morrison and members of the federal Cabinet, September 2021

MINERS, IN TESLAS

We think of superpowers as scientific excellence, foreign-policy nous or sporting prowess, but sometimes the only thing you need is a bucketload of passion and a simple idea. The first episode of *Coal Miners Driving Teslas*, brainchild of climate activist Dan Bleakley, shows us how. Metallica's 'Enter Sandman' is pumping through the surround-sound stereo of the Tesla Model 3 and Mark, a coalminer from central Queensland, wraparound sunnies resting on his forehead, is in the driver's seat. A voice from behind the camera instructs him to slow down before commanding: 'Now just plant it as hard as you can.' The car takes off like a bat out of hell, its speed arriving at 100 km/hour as quickly as Mark's eyebrows arrive at the top of his forehead. The car erupts with swearing and laughter and the camera quickly swings around to reveal a bunch of miners crammed into the back seat, high-vis gear and all. And Mark? He's beaming down the barrel, his mind blown by his first experience of the near future of our transport system.

Long before Mark settled into the driver's seat, Dan Bleakley was growing up in Clermont, Queensland, about 300 kilometres west of Mackay, and a hop, skip and a jump from the site of that climate horror story, the Carmichael coalmine, owned by Adani (now renamed Bravus). Clermont, Bleakley said, was a great place for a kid to grow up, full of open space and hardly a rule applied: 'We would finish school, get a bunch of mates get together and ride our bikes out on dirt roads to creeks, put up rope swings in trees, swing out and do backflips into the water,' he said. 'We'd ride motorbikes and get up to all sorts of mischief out in the bush. We had an incredible freedom out there.'

Bleakley's interest in climate change was piqued on an exchange trip to Germany during his university degree that showed him the contrast between there and here when it came to the problem and potential solutions that were available. But his early career wanderings around the globe meant Bleakley put the issue on the backburner, doing jobs developing machinery in the oil and mining industries, switching life in Queensland for Aberdeen, Scotland, and Perth, Western Australia, before settling in Melbourne to start a printing business with his brother. 'At that stage, I was following the money. I was aware of climate, but I still thought of it as something that was 100 years away, and that governments would ultimately fix it, and it wasn't something I needed to worry about,' he said.

When Bleakley hit his thirties, the growing realisation of the devastating impacts of climate change and slow pace of action lit a fire that could not be controlled. 'It became harder and harder for me to focus on my business when my heart and my mind were really on climate,' Bleakley said. 'I was spending more and more time on Twitter or on Facebook, or reading articles about climate, and less time on the business. We're a small business; we weren't making a lot of money, but ultimately my brother started taking over the business and that freed me up to become a full-time activist.' Bleakley looked to Extinction Rebellion (XR), the UK-born decentralised, nonpartisan global climate action movement where people take different forms of non-violent direct action and other forms of civil disobedience, from family-oriented picnics and large protests often dubbed 'rebellions' to large groups of activists doing 'die-ins' (lying down) in public places, or sitting down in busy city intersections to stop traffic. XR began gaining traction in Australia and in 2019 Bleakley placed himself at the tip of the spear, conducting some pretty hardcore actions, including going on a ten-day hunger strike on the steps of Victoria's Parliament House and super-gluing his hand to the doors of Siemens Mobility, an Adani contractor—that one got him arrested. Around that time, Bleakley took ownership of a Tesla Model 3, the pandemic hit and XR activists looked around at communities locked into their homes, wondering how they could continue to be active on climate.

Bleakley spent twelve months raging on Twitter and Facebook seven days a week until a question from the partner of one of his old university mates made him pause his social media activism. 'She said, "Dan, you've told us how bad it is now, what's the solution?", and I realised I hadn't talked about any of the solutions,' Bleakley said.

Then I thought, I've got a bloody Tesla in the garage and it's awesome: maybe I should use that to articulate what the future could be like. After being on Twitter for two years and watching the way Scott Morrison and certain elements of the media tried to divide us into coalmining communities and people who want action on climate change and constantly trying to split us, I thought this technology could bring us together and decided to show that action on climate is going to benefit everyone and improve everyone's quality of life.

A reprieve from lockdowns and border closures in April 2021 meant Bleakley could make it home to Clermont for a visit, ostensibly to turn up at his twin nieces' birthday party. 'I hadn't been home for years, and when I got to my hometown, I'm in this amazing high-performance Tesla. People had never seen an electric car before, I was quite the show in town.' There were plenty of moments of tension, like at the pub where locals asked the out-and-out climate activist pointedly how long he'd be staying in town. 'It was tricky, I'd go to the shops and some people that I've known for years, they wouldn't even speak to me.' Another of Bleakley's brothers, a worker at a coalmine an hour's drive from town, had been spruiking the Tesla to his mates, and asked to take it there to show it off for a week. 'Just film their reactions,' Bleakley told him. 'My brother put them straight into the driver's seat and he didn't tell him how fast it was, he didn't tell them it's faster than a Lamborghini or anything like that.' With that first clip, aired on YouTube, *Coal Miners Driving Teslas* was off and racing, building a global following, and more and more miners, petrolheads and politicians of all stripes jumped in the machine.

Bleakley has since travelled up and down Australia's eastern seaboard, including taking the ferry to Tasmania, showing people how the world is changing by offering a drive to anyone who's interested: global woodchop champion David Foster went for a drive from his base on the island state, which landed a story on the front page of the local newspaper, *The Advocate*. Bleakley said, 'I've had hundreds of people who have driven this car, and that would have generated thousands of conversations in their own spheres.' (I've had a go and I can tell you, this theory of change is on the money: I've been telling *everyone* how fun this car is and my bloke is an absolute convert.)

Bleakley has learned that all confrontation is the beginning of the fight for climate, but it's not where it ends. 'The coal, oil and gas industries have captured our government and have captured our democracy really. If we acknowledge that, then we know that their interest is in prolonging the fossil fuel industry—and how do they do that? They do that by dividing us,' he said. 'As activists, we have to elevate the issue of climate but we also have an obligation to try to unify our country. We have to articulate the problem and say we're all in this together, we really are.' The cool thing about *Coal Miners Driving Teslas*, Bleakley said, is that it shows inner-city, latte-drinking, avocado toast–eating greenies that miners are awesome people. 'They are really funny, down-to-earth people and we shouldn't be afraid of them. We should just get out there and talk to them about this.'

CHANGE AGENT

It's one of the more striking examples, but Bleakley is someone who has realised his agency in solving climate change. Agency is defined as the capacity, condition or state of acting or of exerting power, and is drawn from the Latin noun *agentem*, meaning 'effective' or 'powerful', the root term *ag-* meaning to drive or move. If you're feeling overwhelmed about climate change and how you can make a difference, it's likely you're suffering from agency deficiency. It's an understandable consequence of having multitudes of choices you can or should make and no shortage of commentary available to show you why the choices you end up making aren't nearly enough. But, as Paul Hawken reminded us, there shouldn't be a debate about whether individual behaviour or government policy is the key to solving this crisis: instead we need every sector of society, top to bottom, to be involved. It *all* matters. The scary thing is, Hawken wrote, that no one is going to save the world on our behalf, but on the flip side, the exciting thing is that every single one of us has a unique role to play and an unparalleled contribution we can make. Like Aussie Rules football, most teams are organised into positions, based on the strengths and talents of individual players, the environment they're playing in and the skills of the opponents they are aiming to beat on the field. Hawken wrote: '*Thinking* you are an individual is self-identity. *Being* an individual is an ongoing, functional, and intimate connection to the human and living world. When we look at our networks, each of us is multitudes. We have different skills and potential, including sharing, electing, demonstrating, teaching, conserving, and diverse means of helping leaders, cities, companies, neighbours, co-workers and governments become aware and able to act.'

The 'hero's journey' is the most established storytelling format we know in our modern world; from when Homer sat down to tell the ancient Greeks about Ulysses to the moment when J.K. Rowling put the final full stop in the Harry Potter series, countless books and movies have followed someone who, after struggle, sacrifice and self-doubt, wins the day. Perhaps it's all these stories that have entrenched the idea that we can only get things done if we're being led by one of these characters who experience this transformation. Carol Sanford, whom we encountered back in Chapter 6, reckoned evolving society requires the opposite of the myths we are absorbing every day: we need non-heroic contributions. 'Non-heroic undertakings need not be grandiose to make a profound difference,' she wrote. 'What they require is an ability to see how the work we are doing—any work we are doing—can play a critical role within society.'

Sanford argued that agency is a core aspect of being human. 'We all want to have an effect on our worlds and to change what we see needs changing. When these avenues of action and contribution are closed to us, we deflate, and our energies curdle into passivity, resentment and self-defeat.' Sanford recommended understanding the unique essence of any entity—so what's yours? Not everyone can kick a footy, owns a Tesla or is interested in going full-time with XR, so let's work on how you claim your agency, wherever you are and at any age.

Dr Ayana Elizabeth Johnson, marine biologist, policy expert and founder of The All We Can Save Project, wrote of how to work out what you should do, by taking a careful look at the skills you have, what brings you joy and what work needs doing, and see where the areas overlap. Inspired by Dr Johnson's approach and drawing from years of helping friends and colleagues work out how to direct their energy, professionally or personally, I've come up with a process that might help you get started or ratchet things up. I'm naming this your Climate Action Awesome Plan, mainly because I'm in my mid-forties and I like cringe-worthy headlines as much as my partner likes telling eye-rolling dad jokes to our teens. Grab a piece of paper, and divide it into four (see

FIGURE 11.1 CLIMATE ACTION AWESOME PLAN: PART 1

Your strengths	Your buzz
'I'm great at ...' *'I can contribute ...'*	*'I love doing ...'* *'I'm excited about ...'*
Skills, knowledge, what others ask me to do and the resources I have at my disposal.	**Projects, activities or experiences I've enjoyed and what you want to learn or experience.**
Think about work or career skills and qualifications, things you know a lot about, and things people keep calling on you to do. Also think about the resources you can contribute, like time or money. These are your unique strengths.	Identify things that don't feel like work when you're doing them, during which, when you're in the zone, time seems to pass too quickly. Add in new subjects, projects you are excited to learn more about.

Figure 11.1): we're going to start building your plan by looking at your strengths and passions first.

When was the last time you reflected on the skills you have, experiences and knowledge you've built, and the talents you are lucky enough to have at your fingertips? What you have right in front of you is the best place for us to start when building our agency. Are you an amazing cook, gardener or accountant? Perhaps you're the person who's great at remembering facts, or the one everyone asks to pull together a family weekend away: maybe you're a compelling public speaker. Can you play the guitar, get a three-year-old to eat a healthy dinner or organise a stand-out party for 1000 people? Are you the person others have on speed dial when they are distressed, confused, angry or need advice? Write it down. In the top left-hand corner, list all the things you know you're good at, what you know a lot about, and what other people are always asking you to do, because chances are you're great at those things too. (No one has to see the list, so you don't need to worry about 'big-noting' yourself. Trust me, you'll be pleasantly surprised by what you see.) Then in the top right-hand corner, write down what you absolutely love doing: the things that are pleasurable, exciting, motivating or just plain enjoyable, whether it's writing detailed foreign policy submissions, kicking a footy, driving fast cars with strangers, making music or pulling weeds from the creek down the road. 'Don't leave out joy!' Dr Johnson wrote. 'For this is the work of our lifetimes, and it need not be an endless slog.'

Paul Hawken reminded us that the most complex and radical technologies are our hearts, heads and minds. I find thinking about what you already have is more motivating than thinking you need to become an expert in climate science or begin training to sail around the world solo, unless of course that's where you want to direct your energy. Nicola Barr recommended connecting climate action to the life you're already in, so what better way to begin than looking at the skills, knowledge, talents and passion you already have that you can direct to our mission? Look at your emerging Climate Action Awesome Plan and be proud! It's likely that you'll end up with longer lists than you had expected in the top two boxes on your page. You're probably already seeing connection points, overlaps and possibilities emerge. For now, kick back and celebrate your strengths: in the next chapter we'll look at how you can direct your energy to generate meaningful climate action.

TOGETHER WE CAN . . . *find our agency*

✳ Every human has agency in the mission for climate solutions, and discovering your agency is the key to managing overwhelming emotions.

✳ You've got knowledge, skills or passions? Yes, you do. Write them down and be ready to deploy them through your Climate Action Awesome Plan.

✳ Remember you're in a team when it comes to climate action: there are many millions of Australians who care about this issue. You don't need to do everything yourself to make a big impact; consider what role you can play with the skills, knowledge and interests you already have.

FURTHER READING

✳ AFL Players for Climate Action, n.d., viewed 21 February 2022:
www.aflp4ca.org.au

✳ Australian Diplomats for Climate Action Now, n.d., viewed 21 February 2022:
www.diplomatsforclimate.org

✳ *Coal Miners Driving Teslas*, n.d., viewed 21 February 2022:
www.youtube.com/channel/UCy0JcHf7dSYHlsnUNDgmTww

✳ Extinction Rebellion, 2022–, viewed 21 February 2022:
https://ausrebellion.earth

* The All We Can Save Project, n.d., viewed 19 March 2022:
  www.allwecansave.earth
* The Cool Down, n.d., viewed 21 February 2022:
  www.thecooldown.com.au

CHAPTER 12

THE BEST TIME TO START

'It's surely our responsibility to do everything within our power to create a planet that provides a home not just for us, but for all life on Earth.'

—SIR DAVID ATTENBOROUGH

Allen Hyde remembers the drive to his parents-in-law's Mount Irvine home back in the 1970s like it was yesterday. The volunteer firefighter and retired engineer painted such a vivid picture over the phone that my imagination was quick to wander back in time to a place that felt unrecognisable, when basalt rainforest covered the far reaches of the Blue Mountains in New South Wales, on Darug and Gundungurra Country. Back then, roads snaking up the mountain carved tunnels through the cooling forest; the canopy was crowded with towering lilli pilli, sassafras, coachwood, Australian red cedar, mountain ash and turpentine.

Retirement meant a permanent shift from Sydney to Mount Irvine, where Hyde is pursuing a multi-generational mission to repopulate the rainforest with these giants. 'When we walk into the rainforest on our property, you feel the moisture just hit your face, even when it's a blistering day outside,' Hyde said. 'You also feel the dramatic temperature drop; it is a wonderful thing.' But the landscape has seen a century of logging and clearing that left the country parched, and when the Mount Wilson Road back-burn (which fed into the Grose Valley fire) took hold in mid-December 2019, over some 53 days many of the remnant scraps of the mountain's rainforest burned, part of the Black Summer of devastation that impacted 80 per cent of the World Heritage–listed region. The rainforests affected have struggled to recover: the land silenced, the fireflies that once swirled around a summer evening all but gone. Hyde, a long-time scouting leader, is in his early seventies, a time of life that is usually more about relaxation than racing around. But after experiencing the fires up close as a Rural Fire Service firefighter, Hyde decided this tragedy must not happen again, and so he turned to nature to help form a solution for the flora, fauna and people too. 'The idea first of all is to improve the rainforest cover itself, and secondly to ensure the protection of our communities,' he said. 'We know from what happened last time, and in other times, if there's enough moisture in the ground and in the rainforest, it is the best protection against fire that we can possibly have. If we've got rainforest around we won't have to burn every couple of years, because we will have more natural protection.'

In 2020, with a small grant from the Blue Mountains City Council, the Rainforest Conservancy was incorporated, with a mission to restore the rainforest that neither Hyde nor his children will ever see grown to maturity firsthand. 'The big thing is the time scale: we are talking hundreds of years to achieve what we're setting out to do,' he said. 'None of the people involved today will see the final product but we have a clarity of what that final outcome will be. We want to see magnificent rainforests in hundreds of years.' The project is starting small, on Mount Irvine, with a call for locals to dedicate 10 per cent of their properties to reforest with rainforest species: five landowners have made the pledge when Hyde and I catch up. The idea is that these pledges will eventually connect to form corridors for wildlife and stronger protections from fires that will without question return to the area. Initial work involves documenting the status of existing remnant rainforest, identifying plant species to formulate which species are prolific, endangered or can be reintroduced from other areas if already extinct or in short supply. 'If we cannot source suitable stock for replanting, we'll propagate stock from existing samples,' Hyde said, noting the success of a trial propagation of 200 lilli pilli trees from a donor tree on the mountain.

A local conservation effort could be where this story ends, but Hyde has a much bigger vision for what this could become. Taking a lesson from the global Scouts movement of around 50 million people, Hyde is setting up the Rainforest Conservancy to be replicable, easily scaled up and self-sustaining, so communities around Australia and the world can pick up the idea and run with it where they are, with central support provided by the organisation in analysis and species selection. Community connection is central to the model—in another hat-tip to scouting—with the aim to get people out onto the land to understand, then help restore the rainforest. 'You do have to understand what you're trying to achieve, why you're embarking on a fairly massive task, because it becomes all-consuming in the end,' Hyde said. 'You need a clear vision, a mission and an objective: all the discipline required to starting a small business applies to this sort of effort.'

When it comes to starting new ventures, the experts say to start with the problem, or 'the pain' you're trying to resolve, and it's obvious Allen Hyde's passion-fuelled project started with something enormously personal. Usman Iftikhar, from migrant and refugee start-up accelerator Catalysr recommended against becoming a ready-made solution looking for a problem: nail the problem you want to solve first. 'It's ideal if it's a problem you are personally facing, and spending more time understanding the problem, interviewing people and learning more about it is really useful,' he said. It's advice we can apply to our efforts on climate, no matter what they are: gather up your insights and get cracking, before you burrow down too many wombat holes. But there is a challenge here: climate change is bigger than every-thing—it touches every dimension of our lives—so how can we make effective action something more tangible and achievable from our own vantage point?

TEST AND LEARN

Transforming our own lives to be more climate-friendly can seem daunting enough, but how do you shift a whole institution? Corinda State High, Australia's first certified carbon-neutral school, is a sustainability story of experimentation and engagement that can teach us a few lessons on how to crank up our climate action. Nestled next to Oxley Creek, 12 kilometres south of Meeanjin, or Brisbane on Turrbal and Jagera Country, Corinda could be mistaken for a normal comprehensive high school with an agri-cultural focus. The sizeable school of 2100 Year 7 to 12 students has the three Rs covered, along with arts, music, dance, science and sporting programs and the bonus addition of an integrated farm on site. What sets this school apart is a pledge made in

2008 to promote environmental sustainability and the fostering of environmental stewardship as a strategic priority. 'One of our missions has always been for students to be stewards of the land, we've had that focus from a long way back,' the executive principal, Helen Jamieson, said. 'I don't do a lot of advocacy around climate change, because I think it speaks for itself with the extreme weather events we're having right now. For me it's about behaviour change and about respecting our land, because Australia is a strong agricultural country.' Jamieson's big pitch is about stewarding the land, respecting it and protecting it for the future.

> The school recognises that it is faced with the challenge to prepare students for rapid economic, environmental and social changes, for jobs that have not been invented and to solve problems that have not been anticipated. Our success will really be measured by the mindset, the culture and the relationships that are lived and experienced through our ongoing priority to create a sustainable future for our students, who one day will be leading the way in their communities.

Corinda's motto is *Hodie quoque cras*, Latin that translates to 'Not only for today but tomorrow also'. The vision for what is required of the school community—'exceed your expectations'—hints at the scale of change that has taken place since the pledge was made in 2008. Massive water tanks capture stormwater runoff that is recycled through an underground irrigation system feeding the sporting fields, oval and school farm, and water usage was halved with the installation of water-saving devices. Solar panels are linked to a monitoring system that students use in their learning. Recycling, controlling weeds and composting food waste, including the local community's coffee grounds, are all priorities on site, where hospitality students are charged with managing garden beds packed with produce that neighbour the classrooms. A fully off-grid, solar-powered double classroom has been installed using funds that would ordinarily have been directed to hiring temporary buildings, and in the last twelve months, 9 per cent of the school's current emissions were avoided by going paperless. There's a fully operational market garden that operates on permaculture principles and helps to feed the 200-odd teachers who work there. It is all monitored and managed with smart technology. Innovation grants foster new projects, like the beekeeping business that one enterprising student got going, which produces between 50 and 60 kilograms of honey per year.

Rob Breuer is co-founder of Zero Positive, a new United Nations–supported program that aims to get all of Australia's 9542 schools to achieve net zero emissions by 2030. We underestimate the outsized contribution schools can make to our nation's overall emissions profile, Breuer said, and he would know, having spent twenty years analysing energy-use data of schools around the country. 'On any given school day, one-quarter of this nation is either at school, working at a school or working in a business that supports a school. It's such a significant part of our society that, if you want to make change, it's where you go,' Breuer said. 'When you aggregate the schools together in energy alone they sit between BHP and Woolworths in terms of their carbon emissions, at 2.2 million tonnes, and when you add in transport, they emit more than BHP.' Zero Positive follows the United Nations Climate Neutral Now pledge that has an approach of learn, measure, reduce, abate where you can—something Breuer said will help schools with the challenges of starting out.

Corinda achieved its carbon neutral accreditation through Climate Active in 2018, ten years after it embarked on its sustainability adventure, with $3000 worth of carbon offsets purchased in 2021 helping it make the grade. This is not where it ends; the school is aiming to achieve carbon neutrality without having to purchase offsets, then move beyond that and, in time, become climate-positive. 'I think you've got to show kids what's possible. You've got to open their eyes so they can see this is what we could be doing and there's more we can do,' Jamieson said. 'We've got to build their capacity to own this, to keep driving it and to be innovative.'

If you're wondering where, when or how to start or scale up your own climate action, before you start your own business, not-for-profit or permaculture paradise, it's useful to consider how entrepreneurs work and think. Masters of start-ups and growing organisations use a *process* of understanding problems or the 'pain', building and testing prototype solutions, iterating and testing again, sharing ideas and building networks along the way. Coach and mentor to emerging climate tech entrepreneurs Mick Liubinskas points me to WD40, a magical degreaser and lubricant that fixes almost anything, whether it's resolving a rusty bike wheel, waterproofing your hiking boots or removing paint smudges from the car: the heady smell of the stuff always reminds me of my grandfather, who couldn't bear the sound of a creaking door. 'Everyone knows WD40, but nobody knows that it stands for water displacement number 40: the inventors made 39 failures on their road to success,' he said. Liubinskas advises start-ups to find their passion areas, look for a set of customers or problems, then really dig in to solve them, which will involve testing sometimes hundreds of ideas because, as he said: 'there's no way your first idea will take off'. Usman agreed. 'Test and test early,' he said, recommending prototyping and engaging people with your solution: sharing your ideas will build networks and open up potential collaborations.

MATCH MAKING

Gretchen Alt-Cooper, an accountant with more than 40 years' tax and superannuation experience under her belt, knew little about cooperatives until she ended up helping to run one in Goulburn, the bustling regional centre of the Southern Tablelands of New South Wales, a meeting place of thirteen Aboriginal language groups. 'I moved to Goulburn twenty-odd years ago from Waverton in Sydney; here I love the clean air, I love the ability to see the seasons change, I love the fact that the city is a community,' she said. 'I feel very engaged with the community.' Alt-Cooper is treasurer of the Goulburn Community Energy Cooperative, a voluntary team that works to use the power of the community to bring power to the community. 'We're in a good location: we've got good clean skies and lots of sunshine, so the community thought, let's harness it,' Alt-Cooper said. 'We would really like to see Goulburn become a renewable energy hub.'

More than 200 members have joined together in shared owner-ship of the investment vehicle that backs the co-op's first endeavour: a 4000-panel solar farm complete with a big battery on a couple of hectares of unused industrial land alongside the railway line. When built, the $4.7 million project will feed enough electricity back into the grid to power around 500 homes and deliver annual invest-ment returns of 5–6 per cent to members. The project, supported by a $2.1 million grant from the New South Wales government, will also set up a community fund, directing at least 5 per cent of overall profits to the local Anglicare branch to establish support services for Goulburn locals experiencing energy poverty, helping with bills and access to energy-efficiency initiatives. Members have contributed more than $2 million towards the project, which has enjoyed enthusiastic community support from day one. 'When we had our first meeting in April 2021 to bring community members up to date, we were about $100,000 shy of our investment target, and by the end of the meeting we had 75 per cent of that pledged. The number of people who are making investments on behalf of their children and grandchildren because they want to be part of this future is phenomenal. We've been utterly humbled by the support from this community.'

Community-owned and -run renewable energy projects are springing up all over Australia (you can see where on the Community Power Agency's online map—details at the end of this chapter), driven by local crews of people who are passionate about energy independence as much as reducing pollution. The Goulburn co-op launched out of connections formed in The Goulburn Group, a sustainability-oriented association of local individuals and groups with the shared view that a transition to a low carbon economy is urgent, locally and nationally. Founded in 2007, The Goulburn Group is an example of a network of strong local ties, united in advocating for social cohesion and action on climate change, economic development and liveability for the city. The network has incubated projects including rehabilitating the Goulburn Wetlands, securing free wi-fi for the city and building one of the earliest electric vehicle–charging stations outside of the big smoke. The energy cooperative is the latest community start-up to bear fruit, with more projects in the development stage, so they can roll out after the solar farm and battery are up and running. 'It's exciting being involved in all of these things. It gives me a bit of hope and makes me feel as though I'm leaving something lasting,' Alt-Cooper said. 'I'm doing something that's making an impact, not just in my own environment, but in a wider sense too.'

Every member of the cooperative's management committee and board is a volunteer, but that doesn't mean just anyone could pull off this complicated community venture. A team was built with deliberation; individuals with climate or sustainability credentials or skills valuable to the organisation were selected to shepherd the process. This leadership group includes local small business owners, information technology and renewable energy sector specialists, and a former local councillor and cattle breeder who uses regenerative practices on his 93-hectare farm just outside town. 'It's important to have people who are philosophically aligned; it makes working together so much easier. We don't have constant debates: we're all on the same page,' Alt-Cooper said, noting the importance of a clarity of vision as much as a diversity of skills. Above all, it's care for each other that helps sustain this work, which is filled with spreadsheets, stakeholder negotiations and investments analysis. 'It comes down to fundamental respect for and caring for each other,' she said. 'Concern for the environment is carrying over into care and concern for each other.'

One of the keys to start-up success, according to Usman Ifikthar, is building your crew. 'Having the right people around you is so important: the people who can actually do the work and advisers and mentors too,' Usman said. But how to go about it? Time to whip out your Climate Action Awesome Plan that you began in the last chapter, and fill in our last two sections. In the bottom left-hand corner, brainstorm the networks you already have access to, which is often the best place to start because you have stronger ties you can leverage. Think about your family, friends, work, local community groups you may be part of or know a member of. The more you list your networks, the more you might see overlapping networks emerge, which could be potential 'wide bridges' (see Chapter 10) in your social networks that could help you to influence change.

Once you've unpacked your connections and networks, it's time to move to the bottom right-hand corner. Start listing what's needed near you or in your 'communities of interest'. Is there a vacant block of land that could be a community garden? Is food waste a problem? Are your work colleagues talking about climate change but not doing much yet? Has your professional association or sporting club pledged to reach net zero emissions asap? Is anyone in your communities or networks talking about climate change, at all? You may need to jump online and see what's already happening or what progress has been made to date. You might list two things here, a dozen or one hundred. Look for what's missing, or what needs more help.

You should finish with a page full of words and before you file it in the overwhelm pile (that's the shelf in my hallway where I chuck the quarterly bills and school permission notes), take a breath and carefully read it over, with a highlighter or coloured pencil in hand. Look at this landscape of strengths, excitement, networks and needs and ask yourself two key questions:

1 Where do my strengths and interests connect to my existing networks and what's needed now? *Draw connecting lines if this is helpful.*
2 What are the top three things I am most excited about when I look at 'What's Required', on the bottom right? *Highlight or circle what's most appealing.*

Your answers should be enough to know where to send your first email or who to call.

FIGURE 12.1 CLIMATE ACTION AWESOME PLAN: PART 2

Your strengths	Your buzz
'I'm great at . . .' *'I can contribute . . .'*	*'I love doing . . .'* *'I'm excited about . . .'*
Skills, knowledge, what others ask me to do and the resources I have at my disposal. Think about work or career skills and qualifications, things you know a lot about, and things people keep calling on you to do. Also think about the resources you can contribute, like time or money. These are your unique strengths.	**Projects, activities or experiences I've enjoyed and what you want to learn or experience.** Identify things that don't feel like work when you're doing them, during which, when you're in the zone, time seems to pass too quickly. Add in new subjects, projects you are excited to learn more about.
Your networks *'Who do I know?'* *'Who seems interested?'*	**What's required** *'What's going on?'* *'What's needed?'*
Connections, workspaces and volunteer groups in your existing circles Name the ties you already have through your local community, or your community of common interest, like your workplace, school or university. Think about people who have other good networks that could be wide bridges that might help ideas or initiatives spread.	**Projects you think could do with a boost or are crying out for attention** List climate initiatives (think energy, transport, food systems or ecosystem restoration) that are missing in your local community or community of interest. Consider what's already around that you could join or help to build up.

In the process of writing this book, I've discovered there are thousands upon thousands of climate and sustainability groups, activities and campaigns up and running all over the country. It's exciting and energising, but leaping into lots of activities and groups can quickly lead to burnout, which is the last thing we need to combat the climate crisis. Before you replace your climate worry with choice anxiety, bear in mind a lesson from movement ecologist Katy Bowman. Bowman, a bio-mechanist and science communicator, has come up with a unique approach she dubs 'stacking your life': it's a concept that involves searching for fewer tasks that meet multiple needs, mimicking how nature goes about meeting its requirements with focus and efficiency. It's a welcome throwdown challenge to the modern-day scourge of multi-tasking, something I believe harms brains, bodies and emotions as we jam in as many to-dos as possible into one timeslot. We need alignment to maximise engagement and possibility. 'Once you identify your needs and which tasks best serve you, you can attend to, pay attention to, get involved in, and focus upon a single task at hand that serves multiple obligations,' she wrote. I agree with Bowman: it's a pretty sweet way of being selective about the way we spend our time.

You've brainstormed away, found some points of energy and perhaps stacked together areas into single potential activities. In the next chapter we'll look at how you can maintain motivation on your climate mission, but, if you are still on the fence about getting your Climate Action Awesome Plan going in the first place, it's time to face your fears and dive into our next story.

PADDLE OUT

For Belinda Baggs, oceans are the great leveller, stripping away social status, race, religion and the rest every time she crosses the shoreline, a single-fin longboard keeping her afloat. 'When you're out surfing, you do realise you're definitely not at the top of the pyramid,' the co-founder of Surfers for Climate said. 'You're extremely vulnerable to other animals, ocean currents, the power of the waves, the rocks and every element that nature can throw at you. All you have to do is keep your head above water and fly with it.' As soon as Baggs could paddle out beyond the breakers at fifteen years old, an obsession with protecting the ocean caught her, as powerful as the impressive swell off Newcastle's coast where she grew up. School days were spent collecting litter during her lunchbreak to stop plastics pollution drifting into the sea, the first signs of what has become a full-time pursuit to stand up for our oceans. 'The more I learned about climate change, the more I realised it wasn't just about plastic pollution anymore, and that it was a lot deeper than that,' she said. 'Over time I realised that the problem was bigger than the individual, that it was a systemic problem, a problem with our politics and a problem with our big businesses. I realised that my time was better used trying to get to the root cause of the problem, rather than always putting bandaids on things.'

Baggs and Johnny Abegg, a former aspiring pro surfer based in Byron Bay on Arakwal Country, founded Surfers for Climate in 2019 after attending a summit on Heron Island in Sea Country, traditional Gooreng Gooreng, Gurang, Bailai and Taribelang Bunda Country, where they learned the scientific backing for the extent of the threat to the world's oceans, and how critical a role the oceans play in the earth's climate systems. The pair was inspired by paddle outs, joining groups of up to 4000 people taking to coastal waters protesting against oil exploration plans in the Great Australian Bight, an effort from a massive campaign alliance that resulted in an inspiring victory. 'It felt more like a celebration when we're out in the water together, like we were celebrating the one thing that we all love the most, which is the ocean and waves and, you know, then trying to do everything that we could to try and save it.' Building a 'sea-roots' movement of surfers began in 2019 and, a couple of years on, thousands have joined up; the governing board boasts members including Holly Rankin (aka musician Jack River), venture capital fund founder Dan Fitzgerald and Adrian 'Ace' Buchanan, a fourteen-year member of the elite World Surfing League professional tour. 'Surfers are from all different walks of life: some of us are sleeping in our vans or on couches and others are living in $20 million mansions on the beach,' Baggs said. 'So we're making sure we can provide a big array of climate actions that every surfer can take, so everybody is making a difference.' When Baggs and I connected, the group was fighting an offshore petroleum exploration acreage in the Otway Basin, directly behind the iconic Twelve Apostles Marine Park off the Victorian coast, and they had also launched a Sustainable Supply Club, whose members enjoy discounts to climate-friendly products including wetsuits and eco–board wax.

These days Baggs, now based in Torquay on Wadawurrung Country on Victoria's Surf Coast, works round the clock on climate and environmental campaigns as a global sports activist with Patagonia and as an ambassador for the campaign to remove plastics pollution, Take 3 for the Sea. You wouldn't know it now, but there was a time when Baggs was holding herself back from full-on climate action, looking to others to speak out.

I felt like I was always waiting for somebody to follow, someone who was going to step in and start talking about climate change. Then I realised maybe the one I was waiting for was me, and maybe other people are in that same situation where they're hoping the same thing. So I decided I was going to, for lack of a better word, put my big-girl pants on and just do it. I'm going to learn as much as I possibly can so I'm confident in what I'm talking about, and if people disagree with me or think that I'm an idiot, that's okay. I don't give a shit, because I believe wholeheartedly, 100 per cent, in trying to stop climate change and save our oceans.

What do experts advise about starting a new thing? 'The biggest thing that I've observed is that the hardest thing to do is to start,' Usman said. 'A lot of people get stuck in this sort of "paralysis by analysis" stage where they'll keep on planning, planning, planning, but not doing anything. It's important to just get started.' It's true that getting started on something new, talking to a new group of people or changing up our routines can feel really daunting, but no one has ever developed without having to face the odd fear or challenge. I'm a big extrovert, but walking into a room of people I've never met, or making a phone call to a stranger is hard to do. I do find that sitting quietly in a new space, getting to know the people over a few gatherings, and not expecting every answer yesterday is far easier than sitting home alone, worrying about the climate crisis.

Liz Wade from the Good Grief Network, whom we encountered in Chapter 2, added that 'it can take time to find your place, and even when things don't go how you expect, it can be a learning experience'. She said:

It's important to follow your heart, follow your intuition, follow what feels right. Follow your own deepest knowing about what is yours to do. The right action for you may sometimes feel hard and challenging, but it will still feel right, it will feel fulfilling. You will feel supported, part of a team that you fit with, and that you are 110 per cent behind the goal of the action.'

Humans of Kangaroo Island's Sabrina Davis, whom we met in Chapter 4, got going with her project of connecting with her community one-on-one because the idea wouldn't leave her alone. 'My head was so full, I never really even thought about fears, I just ran for the hills with it,' she said, looking back. 'My biggest lesson is that if you have a strong feeling about something, it's probably because it aligns with your personality or your skill set so you should follow that up.' Belinda Baggs' number-one piece of advice is not to be so scared, and she's right: what exactly is the worst thing that could happen? Maybe the perfect idea you've been waiting for is the one that keeps popping into your head when you're trying to sleep or tick off the daily to-do list. Maybe the person you've never met is waiting for your call. Maybe, just maybe, the person you're waiting for . . . is you.

TOGETHER WE CAN . . . *start now*

* Build or find your tribe: look at the networks you have and the needs of your local community or community or interest to complete your Climate Action Awesome Plan to get a clue or three on where to start.

* Still feeling overwhelmed? Try cutting the issue of climate change down into chunks that are more manageable—it's what advocates do all the time to find the best place to begin. If you can see a clear line of sight back to reducing emissions, starting where you are, with what you have, this is the best way to get moving.

* Get out of your own way! If you're frozen by indecision or caught up in endless planning, it's time to call up, show up or start your own venture. What's the worst that could happen?

FURTHER READING

✳ Bowman, Katy, *Movement Matters: Essays on movement science, movement ecology, and the nature of movement*, Propriometrics Press, Sequim, Washington, 2016

✳ Climate Active, *Australia's Collective Action*, 2019–, viewed 21 February 2022:
www.climateactive.org.au

✳ Community Power Agency, 'Community energy map', *Community Power Agency*, 2019–, viewed 21 February 2022:
https://cpagency.org.au/resources-2/map

✳ Goulburn Community Energy Cooperative, n.d., viewed 21 February 2022:
https://goulburnsolarfarm.com.au/

✳ The Goulburn Group, n.d., viewed 21 February 2022:
www.goulburngroup.com.au

✳ Great Australian Bight Alliance, *Fight for the Bight*, 2017–, viewed 21 February 2022:
www.fightforthebight.org.au

✳ Rainforest Conservancy:
https://rainforestconservancy.com.au

✳ State of Queensland (Department of Education), *Corinda State High School*, 2021–, viewed 21 February 2022:
https://corindashs.eq.edu.au

✳ Surfers for Climate, n.d., viewed 21 February 2022:
https://surfersforclimate.org.au

✳ Take 3 for the Sea, 2018–, viewed 21 February 2022:
www.take3.org

✳ United Nations, 'Climate neutral now', *United Nations Climate Change*, 2022, viewed 21 February 2022:
https://unfccc.int/climate-action/climate-neutral-now

✳ Your Food Collective, 2022, viewed 21 February 2022:
https://yourfoodcollective.com

✳ Zero for Schools, *Zero Positive*, 2021–, viewed 21 February 2022:
www.zeropositive.org

CHAPTER 13

TESTS OF ENDURANCE

'I am no longer accepting the things I cannot change.
I am changing the things I cannot accept.'

—ANGELA DAVIS

January 2020 was a scorcher for Australia: the year opened with average temperatures at 1.45 degrees Celsius above average for the month. Cities around the nation sweated through those early days of the new decade, but it was Penrith in Sydney's western suburbs on Darug Country that bore the brunt. On 4 January the mercury reached 48.9 degrees—making it one of the hottest places on earth that day. It won't be the last time intense conditions envelop the heart of Australia's largest city, with the CSIRO and Bureau of Meteorology estimating the average number of days over 35 degrees in Western Sydney could increase by *up to five times* by 2090. Rising summertime temperatures have become such a concern that since 2019 all 33 Sydney councils have funded a climate-adaptation program that identifies heat as the number-one climate threat to residents. It is difficult to wrap your head around the pervasive and cascading physical, economic and social impacts more frequent hot days and heatwaves will deliver without more action to support the 2.5 million people who will be affected. That's why I'm in conversation with the community leaders of Voices for Power, a Western Sydney–based organisation. We're talking justice, equity, democracy, community and climate impacts, which are ever-present in a place where temperatures are up to 10 degrees hotter than the city's more affluent coastal fringes.

'It is ironic in a way that affordability comes from how much you can spend on spicing up your house with solar panels, a new boiler and everything that is energy efficient,' said local high school teacher Zubaida Alrubai, Voices for Power spokesperson and co-chair of the group's North-West caucus. 'You do need to have money to save money, and that's not really realistic for a lot of people, especially in Western Sydney. It never is that equitable for the people that are suffering the most.' She added, 'I think equity needs to come from a collective, where sometimes some of the actions that you have to do don't necessarily directly benefit you but then it is for the betterment of the community. It's a "pitch in" mentality.'

Voices for Power, a project of the Sydney Alliance, a coalition of community and religious organisations and ethnic associations, brings together diverse cultural, religious and community leaders to build collective power in support of clean and affordable energy solutions. Sheikh Adid Alrubai, representative of the Grand Mufti of Australia and a devout religious leader of the Muslim communities in Blacktown, walked me through arresting stories from community members he helps, stories of power bills in the thousands of dollars and elderly community members who risk their health when the mercury rises, feeling they cannot switch on the air conditioner because of the oppressive running costs. These are not isolated experiences, with a community survey of 700 people across Western Sydney suburbs, conducted by Sweltering Cities over the summer of 2020–21, finding 55 per cent of people with air conditioners avoid flicking the 'on' switch to save money. 'It's not just equity in a social and economic sense but also in our access to political power,' Voices for Power coordinator and community organiser Thuy Nguyen said. 'When politicians come and talk to our communities or when there's elections, they make promises that they never keep, and people are getting more angry about that.'

In December 2021 the group achieved a big win, with the New South Wales government announcing it would fund installation of solar systems on the rooftops of up to 250,000 homes in the region, with the policy targeting low-income households. The victory, coming after four years of sustained advocacy, demonstrates how the unique approach of deep relationship building across diverse communities is delivering real-world outcomes that are necessary to not only bring down carbon emissions, but also ensure there is justice and equity for people who are looking at the most serious impacts of climate change. 'A lot of young people in our communities want to try to do something about climate change, and Voices for Power is a way to get other young people involved and put their faith in action,' said Ian Epondulan, a youth leader at Catholic Diocese of Parramatta and co-chair of the group's North-West caucus with Zubaida. 'If we can do something now that empowers young people and our local communities, who are not normally heard in the media, to give them a platform to talk about the issues that concern them, I think it's a win for our communities and a win for society.' Under the banner of a Clean and Affordable Energy Roadmap, Voices for Power is campaigning for the state government to implement a 'Healthy and Affordable Homes' policy package that prioritises support for those on low incomes or in rental accommodation. Measures include minimum standards for energy performance of rental properties and grants of up to $5000 based on the outcome of energy audits to install energy efficiency upgrades. The group is also calling for a solar garden pilot to roll out in Western Sydney and a Community Energy Hub to be set up in the region.

What was striking to me over the 90-minute conversation was the depth of common values and generosity of spirit that shone through the group of diverse faiths and backgrounds, testament to the strong relationships built over the years establishing and collaborating on the project. Sheikh Adid was crystal clear on the intersection of Islam and climate change: 'According to my faith, harming the environment is a sin; it is not human. The environment is something Allah gives to us not only for this generation but for every generation, so you have no right to harm it.' Epondulan highlighted Pope Francis's global leadership on climate change and Michal Levy and Sharon Marjenberg, from the Jewish Sustainability Initiative, reminded me that every seven years the faith holds one year as *Shemitah*, a twelve-month period when lands are rested, loans are forgiven, and time is taken to reflect and refocus.

As part of the Voices for Power collaboration, Levy and Marjenberg's group focuses its efforts in Sydney's eastern suburbs, where they partner with local councils to deliver energy efficiency audits and upgrades to people living in social housing, and work with synagogues to run educational webinars on sustainability for members of the local Jewish community. Our conversation revealed deep relationships and the solidarity of common cause that keeps Levy and Marjenberg motivated, with several meetings, joint assemblies and engagement with politicians helping to bridge sprawling Sydney. 'When I heard about Voices for Power, I immediately wanted to be part of this wonderful organisation where I could meet people from all over Sydney from different cultures who we never get to mix with; straightaway we got to work,' Levy said. Solidarity, community and place are the strengths that keep this group powering on, a lesson in connection that we can apply to our own work navigating climate action. Who do you offer your solidarity to? Which relationships are you deepening? It's far from an indulgence to ensure our connections become stronger through our climate journey: in my experience the quality of the relationships I have are fundamental to outcomes in my advocacy work and are a continual source of nourishment in darker moments.

FIND YOUR REUBEN

In 2017 Aussie indie band Cloud Control had racked up achievements that many aspiring musicians dream of: two successful album releases and flying around the world performing live in the United Kingdom, Europe, the United States, India, Hong Kong and Australia. For keyboardist and singer Heidi Lenffer, who had spent a lot of time on those long-haul flights reading up on climate change, a new album in preparation meant it was time to do touring very differently. 'It was the hectic, constantly travelling lifestyle of a musician that gave me the impetus to understand my impact,' Lenffer said. 'We were in the writing process for our third record, in a little Airbnb in Charlotte Bay, on Worimi Country in the stunning Myall Lakes area. We were immersed in this natural landscape and I remember the moment that I sat down and decided to start calling climate scientists.' Surprisingly, many picked up the phone, and she spent half a year having heartbreaking conversations with the experts, an experience that was as emotional as it was educational. 'I remember speaking to a guy who was doing great research into making biofuels from eucalyptus oil, and he was on the verge of tears by the end of the call: I felt like a counsellor,' she recalled. 'It was a sickening sense of dread, to understand the reality of what we're facing.'

Lenffer decided to put her new-found knowledge to work. Following her instincts, she started connecting with her networks and messaging scores of relative strangers in the music industry to found Future Energy Artists (FEAT), an organisation that helps musicians go further than simply paying to offset emissions from time spent on the road. FEAT started out in 2019 by creating an investment fund backing a number of large-scale solar farms, including the 35-megawatt Brigalow solar project in Queensland's Darling Downs on Jagera, Giabal and Jarowair Country. 'It was easier to get people excited about the idea of using their investment to not only have a financial return for themselves, but also build the future that needs to be built: that dual mandate seemed like a no-brainer when I reached out to other artists,' Lenffer said. 'People were really relieved to have someone in their community who was doing something that represented a tangible form of action.' For Lenffer, the project has been a fulfilling way to channel her charged emotions of fear and grief into productive, creative energy. 'If I was still devoting all of my time to writing, recording music and touring, I would just be living with a knot in my stomach through this period, reading IPCC reports and stymieing my creative impulses. I think it's actually enabled me to be a free and influential agent in a critical turning point in history.'

The latest project, Solar Slice, run by sister company FEAT Live, places a small surcharge (recommended at 1.5 per cent) on ticket sales to fund a carbon-reduction program that will invest in solutions for long-term low-carbon live event operations, including mobile solar arrays, purchasing quality Australian-based carbon offsets and setting up clean power purchasing arrangements for venues. When Lenffer and I caught up in early 2022, Aussie band Lime Cordiale had already leapt on board, adding a $1 surcharge to every ticket sold for their planned tour and festival, The Squeeze, created with their managers Michael Chugg and Andrew Stone.

FEAT's stable of supporting artists has grown to include big names like Midnight Oil, Vance Joy, Julia Stone, the Jezebels and Regurgitator, and through the process Lenffer became a corporate authorised representative of a managed investment fund, a big title to add to the résumé. 'I was surprised as anyone because I really never saw myself having anything to do with the finance world, but I think I've got that curiosity and then I just married that up with the need to create something.' The energy sector is notoriously complicated for any newcomer, so I asked her how she stayed motivated over the years conceiving and developing a muso-specific organisation that's all about megawatts and return on investment. Encouraging words from folks in her network of artists made all the difference, Lenffer said, keeping her on the path when the hazards and hurdles of starting something brand new inevitably cropped up. 'If there is someone in your network, and they're doing something that's bloody terrific that speaks to you, then you might be the person they need to help get it up and running,' she said. 'For me, it was Reuben Styles from Peking Duk—the enthusiasm he sent in his texts around investing in solar farms kept me going for maybe four months! Having that validation from a person in a different part of the music community who was as psyched as me was amazing.'

Who is your Reuben? Maybe you need a gaggle of Reubens to keep you going, or perhaps you can be a Reuben to someone in your part of the world who is taking crucial action on climate, helping them to keep up the momentum.

LONG HAUL

If you were to drive north from Sydney to Gloucester, you'd be hard-pressed these days to find evidence that this place was the epicentre of more than a decade's worth of community work to keep coal and coal-seam gas in the ground. 'Younger families have moved here—mainly young women running businesses—there's only one empty shop in the main street and tourists are back,' said Julie Lyford, community campaigner, former local councillor and mayor, and chair of Groundswell Gloucester. 'There's a different feeling now people know the campaign is over.' Gloucester is a picturesque place—think quaint country town—that gently settles at the base of the Bucketts Range (Buccan Buccan), on Worimi and Biripi Country, the gateway to the World Heritage–listed Gondwana Rainforests of Barrington Tops: traditionally timber, beef and dairy country, now horticulture, farming, tourism and small business have been added. Had it not been for the community and dozens of supporters from all over Australia, this place would have seen a coalmine built on the edge of town, and farmland peppered with coal-seam gas wells. What transpired was the nation's biggest carbon polluter, AGL, abandoning its gas plans for the valley and precedent-setting legal work that tipped the mine's proponent, Gloucester Resources Ltd, into the history books.

Searching for the seeds of these successes, we need to go back in space and time to 1970s London. Growing up on the other side of the planet, Lyford was educated about climate change by her sister and cousin who learned about the issue while working at global insurance companies including Lloyd's of London. Arriving in Gloucester in 1986 on the arm of husband Garry, who was taking on the local medical practice, Lyford decided to give her new rural life an added dimension by testing who else in the local community was interested in all things environmental.

> My friend Karen and I used to go walking all the time, and I just said to her 'I'm so worried about this climate stuff', and she said 'well, do something about it, don't just whinge!' So I asked the local newspaper guy to write up an article about a public meeting we organised to see who was interested and he said, 'I'll put it in, but they'll think you're mad.'

Fifty people turned up to that inaugural meeting of the Gloucester Environment Group way back in 1989, decades before concerned community members waged a seven-year fight to fend off energy giant AGL's plans to frack across the region. The campaign involved thousands of people and 32 community groups and organisations standing with Groundswell Gloucester, and in 2016 the champagne finally flowed: the company agreed to relinquish its exploration licence, part of the company's decision to cease exploration for gas in zones across Australia. Had the Gloucester project gone ahead, around 300 gas wells would have been sunk across the beautiful, historic valley floor, some only a few hundred metres from residential areas.

Locals had a few short months to celebrate before plans for Rocky Hill, an open-cut coking coalmine first proposed back in 2009, flared again. The Groundswell Gloucester team suited up once more and in December 2019, the New South Wales Land and Environment Court handed down a judgment that the Environmental Defenders Office (EDO), representing locals in the legal case, described as 'once-in-a-generation'. Chief Justice Brian Preston SC ruled against an appeal by proponents of the Rocky Hill mine, the first time an Australian court had refused consent to a fossil fuel development partly based on emissions and the impacts of climate change. Di Montague was instrumental to the community work that saw the mine defeated, compiling submission after submission to numerous planning processes with support from her partner Chris. How did she keep going over the decade-long fight? 'I'm a bit stubborn, I think; I realised this about myself recently,' she said. 'There were certainly times I wanted to give it up but I was so angry at the developers, mainly because of friends who had been so destroyed by selling their land and that type of thing. I just couldn't believe how unfair it was, and I'm not very big on unfair, so I think that kept me going and it paid off. Thank goodness.'

These days Montague is the vice-president of the Gloucester Environment Group that is still powering on, with more than 100 members and a range of projects and activities on the go. There is the Koala Ways project, which restores corridors for Australia's beloved marsupial and has planted more than 1100 eucalyptus trees; monthly river-care working bees; bushwalking and birdwatching too. Groundswell Gloucester has finished up its work, but new initiatives have been created by the community including Gloucester Transitions, an organisation that unpacks how the region will create a thriving, circular local economy, and Energise Gloucester, which has secured a $460,000 state government grant to build a solar farm in the region. Efforts to keep proposed fossil fuel developments out of town have been successful in part because of long-held community connections formed through groups like this one that are boosted by regular contact, shared vision and a lot of bloody hard work, work that is continually creating new opportunities.

Lyford's advice to anyone worried about climate change is to not get too caught up in the daily minutiae of environmental perfection. 'Don't reinvent the wheel; don't get bogged down in how to do your recycling—that's not what it's about,' she said. 'It's about the pressure from the community for things to change at the political level and the corporate level. There are so many organisations and groups out there that currently exist; find one that you're comfortable with.' She explained, 'My strongest advice would be: don't worry about what anybody else thinks about what you're doing, what you're thinking or what you're feeling, just get out there. Act on your intuition, your gut; there is nothing wrong with always doing the right thing for the planet. Nothing.'

MOTIVATE ME

My teen daughters talk about motivation like it's some sort of tiresome character from the Harry Potter franchise: a ghostly type who hovers in the background before disappearing, nowhere to be found, before a school assignment is due. Writing of the research on motivation, author James Clear said that motivation is not the *cause* of action; it's the *result* of getting yourself moving. In our journey together we've heard dozens of stories that provide insights into how businesses, not-for-profits, climate campaigns and ideas arrive in the world, and what they achieve, but let's consider a few insights that will help you stay the course.

Break out of the bad news vortex: Friends of the Earth's Leigh Ewbank, aka the Vegemite man, said climate change was the mother of all puzzles and recognised that if we don't deliver deep emissions cuts this decade, we're staring at a terrifying future. Does that mean he is across the latest details of emerging scientific research on the scale of the threat? Far from it. 'In my campaigning work, I tend to take a look at the science every now and then, but I don't do it every day. I don't find it helpful.' Instead, he's built a standing item called 'Wins of the Week' into regular Monday meetings of Act On Climate volunteers, something that is energising, allowing positive reflection that keeps the crew sustained. If the goal is to be an engaged person working towards a better world, Ewbank added, we should claim our agency and discern between the activities that will nourish and those that stand to sabotage our change-making efforts. 'I think you need to apply that lens to the choices you're making around the material you're choosing to consume: sometimes reading the *Guardian* newsfeed is not going to be constructive to the work you need to do today.'

Norwegian psychologist and economist Per Espen Stoknes noted that 80 per cent of news on climate change is negative, so it's high time we get things back in balance. In researching this book, I stumbled on a simple hack that could disrupt your doom-scrolling—and you are still allowed to play on Instagram. Next time you're on your favourite social media channel and catch a positive story on climate, environment, community or connection, click the 'save icon' (it usually looks like this 🔖) and pop it in a collection (mine was quite daringly named 'climate hope examples'). When you're feeling like bad news is dragging you down, open your collection and dive in for a read. Not on social media? Sites like Instapaper can serve the same function when you're browsing online. Worried you'll miss the latest update? Friend, in the modern day of the 24-hour news cycle, will you really mourn swapping incremental doom for a side serving of positivity with your lunch?

Make the small things count: AFLW player Nicola Barr is a full-on, full-time elite athlete, so she is the first to recognise that setbacks are always part of the game. In a twelve-month period Barr had to deal with a shoulder dislocation, surgery on her pinky finger, getting hit in the face by a surfboard (another surgery there), and a case of Covid-19. 'A lot of people focus on the really big outcomes, the big goals, but, at the end of the day, that's not the moment when things are just going to change,' she said. 'If you have really good habits around how you can achieve that goal and a good system in place, it's all about the small things that you've been doing along the way. It's about focusing on the things that you can control.' Consider the habits you could be building or consolidating that will mean your Climate Action Awesome Plan is realised. It could mean a regular shift at your local food cooperative, a monthly online meeting with your group of climate-committed professionals, or a reminder on your phone to replace your 10 a.m. phone-scrolling break with a call to that community campaign organiser you need advice from, or to send a check-in message or call out for help to your WhatsApp group full of climate-committed friends. Creating a couple of regular habits that align with bigger goals will make progress feel more automatic, and remove the need to make a decision to do or not do, which is shaky territory for maintaining motivation. As James

Clear wrote, scheduling goals puts decision-making on autopilot, by giving them a time and place to reside, while creating regular rituals removes the need to make decisions. 'Stop waiting for motivation or inspiration to strike you and set a schedule for your habits. This is the difference between professionals and amateurs. Professionals set a schedule and stick to it. Amateurs wait until they feel inspired or motivated.'

I think you know what to do.

Take a breath: Don't worry: this is not where I evangelise that yoga and meditation are a must (given my record on the meditation front, I suspect the sermon would fall pretty flat). But I must and will make the case for rest, and regular breaks. GetUp's First Nations Justice Campaign Director Larissa Baldwin saw friends burning out from relentless advocacy, and found out the hard way herself, working so hard she fell ill. Baldwin said that we need to lose the burn-and-churn culture, which I think equally applies if you're donating your time by volunteering on the side or you're working 24/7 in a climate-connected job. 'There's this heroic thing that's associated with being willing to throw everything at something and I think that we glamorise that, but if everybody does that then there's no one to do the work,' she said. 'If we're really talking about the change that's needed on climate, we need to conceptualise the ways that we work very differently and we need to be more protective of people.' In *Burnout: The secret to unlocking the stress cycle*, sisters Emily and Amelia Nagoski wrote that we are engineered to oscillate between working and resting: when we allow this cycle to happen, our health and our work benefit. The brain must be given regular breaks, they wrote, so it can hum along in the background in what neuroscientists have named 'default mode', restoring us from depleting activities and allowing our motivation to return. 'Walking away from a task or problem doesn't mean you're "quitting" or giving up. It means you're recruiting all your brain's processes for a particular task—including the capabilities that don't involve your effortful attention,' they wrote, and I am cheering them on. By making habits of breaks—daily, weekly and across the year too—you will be replenishing your brain and allowing space for you to enjoy life in all of its dimensions, which is what we're working to protect.

TOGETHER WE CAN . . . *keep going*

* Solidarity and mutual support build resilience across diverse communities, so consider how you can build support for efforts taking place in your local area or your community of interest: it will help build relationships that will nourish you.

* It can take as little as one message of support to keep you going for months, so call out for support from your crew when you need it, and offer support to people doing amazing work in your networks. It helps—a lot.

* Feeling overwhelmed from time to time is inevitable when it comes to climate action, but to stay motivated you need a few hacks to break out of the bad news vortex. Split your goals into smaller, regular habits and rituals, and remember to build in regular breaks too.

FURTHER READING

* Clear, James, 'Motivation: The scientific guide on how to get and stay motivated', n.d., viewed 21 February 2022: https://jamesclear.com/motivation

* Energise Gloucester, n.d., viewed 21 February 2022: www.energisegloucester.org

✳ FEAT.Initiative, *FEAT*, n.d., viewed 21 February 2022:
  www.feat.ltd

✳ Groundswell Gloucester, n.d., viewed 21 February 2022:
  www.groundswellgloucester.com/index.html

✳ Nagoski, Emily & Nagoski, Amelia, *Burnout: The secret to
  unlocking the stress cycle*, Ballantine Books, New York, NY, 2019

✳ Stoknes, Per Espen, 'How to transform apocalypse fatigue into
  action on global warming', *TEDGlobal*, September 2017, viewed
  21 February 2022:
  www.ted.com/talks/per_espen_stoknes_how_to_transform_
  apocalypse_fatigue_into_action_on_global_warming

✳ Sweltering Cities, *Sydney Summer 2020–21 Survey Report*, 2021,
  viewed 21 February 2022:
  https://swelteringcities.org/summer_survey

✳ Sydney Alliance, *Voices for Power*, n.d., viewed 21 February 2022:
  www.sydneyalliance.org.au/voices-for-power

CHAPTER 14

HOW WE NURTURE HOPE

'Together, we can build a remarkable country,
the envy of the rest of the world.'

—DR LOWITJA O'DONOGHUE AC CBE DSG

'When you get to a protest, there's a really strong sense of community and unity between everyone, the air feels really warm,' school striker Ashjayeen Sharif said. 'You can smile at anyone you're walking past and know that they care about this really important thing that you care about as well. It feels really empowering.' Ashjayeen and I were having deep chats, talking courage, corporate challenges and community, all sizeable topics for a spring evening conversation. Ashjayeen's interest in keeping the planet healthy began in primary school, where he learned the other three Rs (reduce, reuse, recycle) and developed his interest in environmental protection and urban planning. A few years later, when his Instagram feed showed him a small teenager with plaits sitting alone outside the Swedish parliament, something clicked.

Inspired and spurred to action, Ashjayeen made a sign and headed to a local Brisbane gathering of the first national school strike in Australia, in November 2018, when around 500 young people turned up. 'I'd never been to a protest before that and I guess I've never really encountered protests apart from what I'd seen in the media. So I went with a couple of friends and that was, like, the turning point. I just immediately thought it was so cool.' Since that first school strike, young people like Ashjayeen have turned up to protests in their tens of thousands, and organised into a robust network of people who are committed, connected and creative, despite all the upheaval from fires, flooding and pandemic lockdowns.

In 2021 Ashjayeen, by now a student at Monash University in Melbourne, took his concerns to the top of the nation's energy sector, running for a vacant seat on the board of AGL, in collaboration with Greenpeace Australia–Pacific. In front of shareholders at the company's annual general meeting that year, he presented a petition signed by 18,000 people and pitched his five-point plan to reform the company's climate credentials, including having the board commit to necessary emissions reduction targets set by the global Paris agreement, close polluting coal plants by 2030 and ensure support for workers through the transition. (Ashjayeen isn't the only one with this idea: in early 2022 tech billionaire Mike Cannon-Brookes had a crack at purchasing AGL in a consortium bid with Canadian pension fund giant Brookfield.) Campaign stunt it was for sure, but the move provided a platform for Ashjayeen to talk publicly about how climate change is affecting people in Bangladesh, home for many members of his family. 'I was lucky enough to be able to move from Bangladesh to Australia, which is better protected from climate change and the effects of the climate crisis, but hundreds of millions of people probably will never be able to. So while fighting for climate change here is fighting for our future, I know that it's also fighting for the present of my people.' Ashjayeen added:

I was worried that my family would tell me to be careful doing big publicity and getting involved in big corporations. It stems from a stigma associated with activism back in my home country, because if you get involved in politics or activism there, it's actually violent and politicians are much more visibly corrupt, but my family was super supportive of it. The greatest thing about it was my parents really felt a sense of pride that I had made the news and everything: my mum and dad put so much effort in, sending the petition to all of their relatives and all of my extended family were really impressed too; it was so cute. It gave me a really strong sense of hope that people from such diverse backgrounds, people who aren't engaged at all in environmentalism, supported the campaign.

Anjali Sharma has a busy life that's very familiar to my household: she's studying her final year at Melbourne's Huntingtower School, and holding down a part-time job at JB Hi-Fi. Friends, fun and huge love from her part-kelpie, part-staffie pooch Maya are part of life as a teenager on the cusp of adulthood. Sharma has a little more on her plate than most, as one of eight young people across multiple states who took on federal Environment Minister Sussan Ley, seeking a court ruling to stop the expansion of the Vickery coal project 25 kilometres north of Gunnedah in northwest New South Wales, on Gomeroi Country. After a four-day hearing for what is now known as the Sharma case, Federal Court Justice Mordecai Bromberg ruled that Minister Ley had a duty of care to avoid harming young people from climate change, a world-first finding that ricocheted around the globe. 'Expert evidence established that emissions from this single mine might be responsible for tipping points that could lead to a terrifying four degrees of warming,' the lawyer who represented the students, David Barden, said. 'The decision legitimised the concerns of a generation of younger people who want to see an end to new coalmines for a better future.' Barden went on: 'The court stated inaction by adults on climate change "might fairly be described as the greatest intergenerational injustice ever inflicted by one generation of humans upon the next".' It was a powerful counterpoint to the unabated development of coalmines in Australia.

While I chatted with Anjali, seventeen, and her legal guardian of sorts (the technical term is 'litigation representative'), retired teacher and Brigidine nun Sister Brigid Arthur, 86, the pair have already spent months patiently waiting for a decision on the Minister's appeal. That's right—an appeal by our government against a ruling that our representatives should look after our kids. It's galling to think about this, but Sharma was upbeat. 'If you were to tell me that, one-and-a-half years ago, I was going to be the litigant in a case that went against the government I would have been like, "You're crazy!" I was just organising strikes and little bushfire vigils and stuff like that, and now here I am.' But Sister Arthur, who has a good seven decades of life experience on Anjali, was unsurprised by what the teenagers have stepped up to do. 'I was a secondary school teacher most of my life and I've watched kids who, when really given a chance, can do all sorts of amazing things,' she said. 'Often we don't give them the chance and these kids have seized their chance. I've certainly been impressed by their passion for doing something that is right and something that is forward-looking and that requires an intelligent response.' She added:

It's just a pathetically immoral stance to take if you've been given the opportunity to lead in whatever part of the government or important areas in society not to use that influence and power to do good. It just leaves me speechless. Climate change is not just an important issue: it's life-and-death stuff. I won't be around, but for future generations, it's absolutely imperative that they do the best that they possibly can.

The government was successful in its appeal, but in its decision of March 2022, the full bench of the Federal Court did not dispute the facts, noting that the risks and dangers of global warming were never in dispute. Outside the court, young people wept and embraced, but Anjali stoically fronted the media with a promise: 'This won't stop us in the fight for climate justice . . . whether we appeal or not, we'll be back.' Elaine Johnson, director of legal strategy at the Environmental Defenders Office, wrote of the legacy of the case, which has built the foundation for legal action by young people, including a challenge by children on human rights grounds over the approval of a coalmine in the Galilee Basin in Queensland. 'In the end, cases are won and lost, that is the nature of the game,' Johnson wrote. 'What's important is the impact these cases can have, the history we are making just by taking them on and having these matters considered by the courts . . . In 50 years' time, nobody will remember which cases were won or lost, they will remember the bravery of the young people who stood in front of the world's courts, fighting against government inaction and complacency on climate change.'

Anjali built climate activism into her life because the cause is intensely personal, her family having emigrated from India when she was only ten months old. 'It's a real sense of survivor's guilt, when I think about how I feel living in Australia while I watch my family in India constantly gear up for terrible summers,' she said. 'I went back in April 2021 and it was literally too hot to leave the house; it would be 40–45 degrees by 9 a.m., it was almost like a joke. We've only just started wearing masks because of Covid, where people in India have been wearing masks for decades just because of the amount of industrial pollution.'

Anjali's family is proud of her work (there's been frenetic activity on the family WhatsApp group at peak moments in the case) and her words lean more to the opportunity she's had to be part of the case than any burden. There's no question that for Anjali, climate action has been too intense in some moments: it's a chilling reminder that the generations to come are going to have to do some heavy lifting. 'Sometimes, it's like, you just want to be a normal teenager, you know what I mean? Sometimes it gets very overwhelming,' she said. 'I'm doing a lot of stuff alongside school, so I do sometimes feel like it's stopped me from being a teenager in some ways, but I feel very grateful that I've had the opportunity to talk about my family story and about my family in India.'

When I asked Anjali what advice she has for all of us, she reminded me that there is a place for everyone to act on climate, no matter how new someone might be to the issue. 'I'd say you're a lot more powerful than you think. Just get yourself educated, get organising: the world won't do it for you.'

The experience of running for the board of one of Australia's biggest companies has helped Ashjayeen build his confidence to stand up and demand justice, to think bigger about what's possible. 'If I was to give someone advice, I would say that, don't be scared of reaching for things that you think are absurd,' he said. 'That saying "reach for the skies" is absolutely appropriate, so set huge goals.'

NEW BEGINNINGS

I'm a relative latecomer to the threat of climate change, so at the time I was ready for babies it didn't cross my mind that I wouldn't have children. My only hopes were that my children would be healthy, happy and develop love for people close and far to their own small circle of experience, and a deep appreciation for the natural world, that's all. If I had to make the choice right now, would I do it? It's a question I thankfully do not have to answer and, TikTok memes and driving lessons aside, I am utterly in love with my daughters: I cannot imagine making a choice that would have excluded the possibility of a life with them in it. These days the two young women I apparently had something to do with rearing are racing towards being all grown up, and I fear what might be required of these two vibrant, feisty souls to thrive. The world feels far more threatening than when they arrived on the earth, so it's little surprise that people arriving at the kid-making age and stage are carefully considering their choices. A 2019 survey of 6500 supporters by the Australian Conservation Foundation (ACF) and 1 Million Women found 33 per cent of women under 30 were reconsidering having children or more children due to concern of an unsafe future brought about by climate change. In 2021, a poll of 15,000 Australians conducted by YouGov and commissioned by the ACF found 20 per cent of people between 18 and 24 years old had reconsidered or chosen not to have children as a response to climate change. Given these numbers, it feels like the very decision to have children now is a choice to be hopeful, which is why I turned to doctors Kate Lardner and Harry Jennens to help me learn more.

Partners in life, love, work and activism, Dr Lardner and Dr Jennens spoke to me from the bunker of the first few months with not one, but two new babies. The relentless grind of those early days and nights with a newborn came flooding back as I peered through the Zoom screen. Their new-found expertise in passing four-month-old twins from one set of arms to another was awe-making—these parenting pros are all over it, despite dummies flying, cot swaps and feeds pausing our conversation. The newborn choreography formed the backdrop to a deep discussion on the decade-long decision-making process that resulted in babies Clio and Quinn arriving in the world. 'There was the emotion of wanting to have children, but I couldn't make peace with it until I learned about building collective power and influencing systems change,' Lardner said. 'I learned that my individual decision not to have a child actually wouldn't really change much, especially when we decarbonise our economy. Knowing that we have the ability to change the systems within which we operate meant I could make peace with the decision to have a child.'

Lardner and Jennens met through volunteer work with Doctors for the Environment Australia, becoming closer during a year pulling together the organisation's annual conference: their first date was ice-cream at Melbourne's Southbank. Jennens, a general practitioner, came to climate change from a global health and justice perspective, set on the path when he heard a climate activist from Bangladesh speak at a Students of Sustainability conference in the second-last year of his lengthy medical degree. 'Someone in the audience asked what we could do to help and he said: "Countries like yours are where all the emissions are coming from, so you need to fix that in order to help us." I was struck by the responsibility that we share here in Australia and in other developed countries and I realised this was the most urgent issue to work on to protect people's health around the world.' Lardner, when on the path to becoming a qualified hospital doctor, initially harboured scepticism of the climate challenge's extent, but three months' research put those views to bed. Scepticism was replaced with paralysing anxiety for a time that, she said, was debilitating. 'I did what most people do, started attending lectures and meeting other people, but that didn't really lessen the anxiety, because I didn't understand how at that point I wasn't really working with other people to make change,' she said. 'But once I learned about strategy, power, and collective action, climate change doesn't bother me now, I'm not anxious about it, I'm hopeful.' Proving her point, she added, 'Addressing climate change is us against the fossil fuel industry, and I think we will win.'

In 2015 the pair founded Healthy Futures, a group in which health workers, students, and community members work together to advocate for emissions reductions on health grounds. 'There are health co-benefits to many interventions that address climate change,' Lardner said. 'Increasing active transport, decreasing red meat consumption, reducing air pollution—all of these measures produce individual health benefits as well as helping address the climate crisis.' The group's actions go straight to the source, with health workers creatively confronting fossil fuel companies and superannuation funds in full scrubs (think blue gowns, masks and caps) outside their corporate offices and agitating at their annual general meetings. In 2020, after a five-year campaign from the group, in partnership with Market Forces, the $52 billion health-industry superannuation fund HESTA dumped thermal coal and committed to getting its investment portfolio to net zero by 2050. The campaign is an example of the systemic change that Lardner and Jennens orient their life and work around, and that Healthy Futures was set up to achieve. 'Despair is a natural reaction—it's logical—but once you understand that climate change is a problem generated largely by corporations and governments that have to change their behaviour, and we have to collaborate to bring to bear our collective influence on them, the path forward becomes clear and purposeful,' Jennens said. 'It's important to appreciate the institutional causes of climate change and work to change them rather than only focusing on individual consumptive behaviour.'

Our conversation had turned decidedly prosaic, which is to be expected when you're in diagnosis mode with two doctors, so I asked about the poetry of creating and nurturing new lives. 'Family is pretty much the most important thing in our lives, and we wanted to emulate what we had with our families by having our own family: joy and love,' Lardner said, 'and if you look at the younger generation today, they are extremely inspirational and they provide hope as well, they teach us.' Jennens agreed that joy is something we need to sustain us and renew our purpose: 'When we're working to create a world in which people can live freely and healthily, it doesn't make sense to me to deny ourselves the opportunities that we're fighting for everyone to have. So, yeah, I think having a couple of kids is okay.'

Marie Carvolth spent over a decade working as a coral ecologist, studying corals and conservation in some of the most remote and beautiful parts of the Indo-Pacific. Awareness of climate change crept up on her over the decades as the science developed and she understood the impacts on her beloved corals, including the Great Barrier Reef—where, at nine years old, snorkelling off the Low Isles, she committed herself to becoming a marine biologist. Carvolth had witnessed coral bleaching events several times over and began her journey into climate advocacy, signing petitions, switching banks and joining the occasional protest, but it was the arrival of her son in 2014 that transformed her action on climate change from optional to essential. 'It all became just so intensely real and personal and urgent,' Carvolth said.

It suddenly became my personal responsibility. There was no choice: I have to fix this for my son to be safe. My son is seven years old and every one of the last seven years has been the hottest on record: that's pretty full on to me, that he's living in these times with the trajectory that we're on. In the year 2030 he'll only be sixteen years old, so we'll have sealed his climate fate one way or another before he even has the right to vote, so it's my parental responsibility. That's why I do what I do; I have the privilege of spending time on this and I can't see any other more important way to spend my time.

Australian Parents For Climate Action launched in early 2019, after Carvolth connected with a small collective of parents, formed a Facebook group, and—with co-founders Suzie Brown, Heidi Edmonds and a handful of other passionate mums—the group blossomed in under two years to 16,000 supporters and 30 local groups across every state and territory. 'Our mission the whole way along has been to engage, empower and amplify parents' voices to advocate for climate action in their communities, in the media and directly to decision-makers, to increase the political will for climate action,' Carvolth said. 'We're fiercely nonpartisan, because it doesn't matter who is in government: if we don't have support from both sides we're not going to make the progress that we need. We've got so many amazing solutions that are available to us—it's so exciting, we just need the political will to implement them.' The group's Solar Our Schools campaign inspired thousands of parents to take their first steps into advocacy, and helped shift over $71 million in funding in three states (Western Australia, New South Wales and Tasmania) to fund solar panels and batteries at schools and childcare centres. 'Politicians and parents alike call it a "no-brainer",' Carvolth said. 'It's a fantastic opportunity to cut energy bills and carbon emissions, and redirect the funds to learning resources, all while helping to create regional jobs.'

We turned over thoughts on what it is to feel hopeful, Carvolth reflecting on how parents have been nurtured through creative campaigning activities, holding 'play in' protests at federal parliament and outside politicians' offices, and finding solidarity, community and connection along the way. 'So many parents have said they were worried and didn't know what to do and are grateful for the support and nurturing in this group. They don't want to feel more terrified and they don't want to be reprimanded for what they're not doing: there's enough "parent fail" and "mum fail" guilt around already.' Hopefulness is a determination, she said, a decision to take actions large and small that all make a difference. 'If we give up on acting on climate, we are giving up on our kids' safety,' she said. 'It's just like if you were told your kid had a serious illness but there is a cure: you'd do everything you could to achieve that cure. That's what gives me the motivation to keep going.'

My friend and advocate, campaigner and inspiring energy nerd, Nicky Ison, was 27 weeks pregnant with her first baby, all second-trimester awkwardness, with her next hospital check-up a couple of days away when we caught up. As Energy Transition Manager at WWF-Australia, Ison is on a mission to make Australia a renewable energy superpower, to make the transition faster *and* fairer for all, but we talked less shop than usual over our Friday afternoon cuppa. 'I think it's incredibly important for people who care about making the world a better place to bring people into the world, to have kids,' she said.

The world has always had terrible things happening to it. Did people ask questions about whether they would have a kid during the rise of fascism? As much as I'm petrified about climate change, I am also optimistic and so I choose to bring that optimism into my life and my family. Also, I don't want the people and institutions causing climate change and continuing to pollute the planet with fossil fuels to stop me and my family doing something that will bring us joy. I don't want to give them that power over my life.

If you've ever witnessed Nicky educating a crowd on the ins and outs of the energy system and making the compelling case for intervening in systems to create outsized change, you might be surprised to learn that Ison is also the first on the dance-floor at any party, and the rest of us can barely keep up. At age seventeen, Ison could have chosen to study contemporary dance, but thankfully took the engineering path, as her way of contributing to climate action. Hope for Ison is material, her engineering brain connecting to the massive rates of renewable energy growth in Australia that as little as five short years ago, the pundits and the politicians said couldn't be done. But the dancer's big, brave and wild heart is ever-present too, as she argued for all of us to hold and nurture the joy we need to sustain ourselves, evoking the sentiment of early twentieth-century feminist activist Emma Goldman. 'I think joy is incredibly important to the success of social movements,' she said. 'That saying "it's not my revolution if I can't dance", is something I've held close for the entire time I've worked on climate. Not only does it link to my dance roots, but it's also a reminder to have fun: fun is an essential part of creating change.'

NO HOPIUM

In the final throes of writing this book (which bears an uncanny resemblance to the final stages of pregnancy: emotional roller-coasters, weight gain, self-doubt and the rest), I asked scores of people in my networks of, for the most part, climate-conscious folks, what made them feel a little more hopeful. Passionate friends, colleagues and collaborators named campaigns and organisations big and small that are making gains every day around the country: some of these stories you have read about in earlier chapters, and you will find many more in Climate Action Starts Here, at the end of this book. Friends highlighted impressive technological innovations; others named people and communities who are making headway in the places near them. Personal acts of gratitude for a sunrise, a bird in flight or the inner peace brought by swimming in the ocean, walking in nature or taking a moment to meditate and quietly reflect also made the list.

This free-forming flow of feel-good was interrupted with a challenge posed in the form of a question: is hope paralysing and pacifying? It's a fair call, if you consider the word is drawn from the Old English *hopian*, a theological virtue that assumes hope for salvation or mercy, or to have trust and confidence in God's word. Emerging in the zeitgeist is the idea of *hopium*, described by some as a drug-like false hope that we turn to in a time of undeniable crisis that manifests as an insidious form of denial. Thinking everything will be fine is, frankly, unrealistic and obvious every time you tune in to the news. In our time together we've unpacked how common it is for people to experience profound anxiety, grief and other charged emotions, but if we allow ourselves to be caught in endless cycles of these emotions we will indeed become paralysed by hopelessness. Science communicator Britt

Wray wrote of how our emotions are trying to tell us something: by approaching our psychological responses to the climate crisis with curiosity, we are able to be more self-aware about how we are (or how we are not) showing up at this moment in history, then adjust accordingly.

> After decades of squandering time, the only choice we have now is to take striking justice-oriented approaches to addressing the crisis at all levels of society, at the same time that we are forced to contend with losses. The point of talking with others about the dread we feel on the inside is not to emotionally reckon ourselves into oblivion. It is to help ourselves cope as well as harness the feelings, so that when we connect with others we can more effectively transform the world on the outside.

Talking through our emotions, with a group like those established by the Good Grief Network, or by connecting with a climate-aware professional can be helpful in developing this awareness of what our purpose is, in this place, at this time in human history. But hope won't fall from the sky from this process alone— we do need an intentional, engaged form of hopefulness, something Joanna Macy and Chris Johnstone described as 'Active Hope'.

> With Active Hope we realize that there are adventures in store, strengths to discover, and comrades to link arms with. Active Hope is a readiness to discover the strengths in ourselves and in others; a readiness to discover the reasons for hope and the occasions for love. A readiness to discover the size and strength of our hearts, our quickness of mind, our steadiness of purpose, our own authority, our love for life, the liveliness of our curiosity, the unsuspected deep well of patience and diligence, the

keenness of our senses, and our capacity to lead. None of these can be discovered in an armchair or without risk.

Emily Ehlers argued that hope is a verb, something that we must actively cultivate. 'What if we nurtured hope like a tiny flame, feeding it with new visions and inspiring possibilities, daily acts of kindness and courage, stronger communities and a deeper sense of purpose, until that flame started to glow on its own?' she asked. Indeed, what if?

To realise this type of hope, we need to think of hope as a *choice* we make and a *practice* we develop over time. First, hope is a *choice* we make to believe there is potential to heal this world. The vast majority of people who have told their stories in these pages are choosing to be determined, to be purposeful, to see the goodness and potential in their fellow humans and act accordingly. Second, hope is a *practice* of building our collective resilience with action and connection, something Paul Hawken recommended. The purpose of this book is to open the door to some inspiration, ideas and frameworks that may be helpful for you to set some goals and strategies for action on climate change that are meaningful, yes, but also influential. If you've made it this far, to be honest I'm a little concerned, because I'd rather you threw it down and acted on climate change; after all, it's up to all of us to create the change we want to see, together. Ask yourself: do you see the clear line of sight between your goals and strategies and the systems you want to influence? Are you harnessing your agency? Can you stack climate action into your life so it's more manageable? Are you taking regular breaks? We've looked at the emerging groups across finance, sport, politics, agriculture, technology and local communities that are the vast expanse of this growing social movement, so there are plenty of places to connect and contribute. If you're overwhelmed or confused by choice, I recommend just picking something and see where the adventure takes you. Remember these words from anti-nuclear campaigner Helen Caldicott: 'Everyone can be extraordinarily effective, they just have to not be self-indulgent or narcissistic or greedy, and work for other people and other things. In that action lie the germs of true happiness. You'll never be happy trying to make yourself happy. It doesn't work.'

TIME TO SHINE

At the top of my street, there is a bushwalk that will get you to the top of the Illawarra Escarpment on Dharawal Country in under 30 minutes if you're in a pure state of mind over matter. It's a grand total of 1000 steps straight up, a fact I know only because one of my daughters counted them a few years back in an effort to take her mind off the climb. The walk is like a shock to the system every time I hike it, no matter if I'm pulling myself off the couch for the first time in three months or if I'm in a phase of hyped-up bootcamp training. It's all too easy to avoid walking this path, or put it off until tomorrow, because I know from experience that every time my legs will burn with fatigue and my lungs will heave with the effort of putting one foot in front of another. But I never regret the climb, because for all its pain it is stunningly beautiful too, the never-ending staircases winding through the ferns and towering eucalypts that stand beside the cabbage palm, before arriving at astounding views of my new home. I take a few minutes to reflect on how remarkable it is that we exist here, before the much quicker journey back home, the endorphin rush powering the rest of my day. Acting on climate change is a monster of a climb, but one that has beauty and wonder in the journey just as much as the destination. So I leave you with a call to love this country in all its pain and history and complexity. Be proud of what we have managed to create, inspire and change for the better, despite all of the damage. Grit your teeth and get to work to heal what's pained and dying in our people, our culture, our natural world. Honour your ancestors, and your children's children through your contribution. Connect with your community and remember to always, always dance.

Love this country. Fight for it.

TOGETHER WE CAN . . . *be purposeful*

* Our emotions are a teacher that guides the way, rather than an enemy to run from or battle. Experts say that we should approach our emotional responses to climate change with curiosity, examine how we are showing up at this moment in time and adjust our actions accordingly.

* Ask yourself: do you see the clear line of sight between your goals and strategies and the systems you want to influence? Are you harnessing your agency? Can you stack climate action into your life so it's more manageable? Today is the perfect time to begin making your contribution.

* Rather than thinking of hope lulling us into a complacent oblivion, we should consider being hopeful as a choice we make to believe there is potential to heal this world, and a practice of building our shared resilience by combining action and connection.

* There is a place for everyone to act on climate, so take a lesson from the students striking for climate justice, and who are making history by challenging the fossil fuel industries in the boardrooms and in the courts. Remember these words of advice from Anjali Sharma and Ashjayeen Sharif: don't be scared to reach for goals that might seem entirely absurd. You're far more powerful than you think you are.

FURTHER READING

* Australian Conservation Foundation, 'Australia's largest ever climate poll is here. See the results', n.d., viewed 21 February 2022:
  www.acf.org.au/climate-poll

* Australian Conservation Foundation and 1 Million Women, *What do Women Think About Climate Change? Summary of a survey of 6500 women by the Australian Conservation Foundation and 1 Million Women*, n.d., viewed 21 February 2022:
  https://d3n8a8pro7vhmx.cloudfront.net/auscon/pages/10649/attachments/original/1549598020/4pp_women_and_climate_change.pdf?1549598020

* Australian Parents for Climate Action, 2021–, viewed 21 February 2022:
  www.ap4ca.org

* Doctors for the Environment Australia, 2020–, viewed 21 February 2022:
  www.dea.org.au

* Ehlers, Emily, *Hope is a Verb: Six steps to radical optimism when the world seems broken*, Murdoch Books, Sydney, 2021

* Healthy Futures, n.d., viewed 21 February 2022:
  www.healthyfutures.net.au

* School Strike 4 Climate Australia, 2022, viewed 21 February 2022:
  www.schoolstrike4climate.com/

* Wray, Britt, 'Talking about dread is not enough—we need action too', Gen Dread, 30 June 2021, viewed 21 February 2022:
  https://gendread.substack.com/p/talking-about-the-dread-is-not-enough

CLIMATE ACTION STARTS HERE

Together we can . . . *transform our lives*
Together we can . . . *understand grief*
Together we can . . . *build momentum*
Together we can . . . *reconnect with each other*
Together we can . . . *find common cause*
Together we can . . . *see what's possible*
Together we can . . . *make it*
Together we can . . . *maximise our impact*
Together we can . . . *reinvigorate our democracy*
Together we can . . . *be influential*
Together we can . . . *find our agency*
Together we can . . . *start now*
Together we can . . . *keep going*
Together we can . . . *be purposeful*

EVERYTHING YOU NEED TO GET MOVING

You're recycling your garbage and signing a petition or two, that's great! Thinking it's time to level up? Here's how you can help get Australia's emissions down in this critical decade. Remember it's okay to change things up if you're involved in something that's not the right match for you. *Every action we take makes a difference.*

For an updated list and more tips, articles and inspiration, head to **climateactionstartshere.com**

SWITCH OUT

Okay, you've heard it before, but have you done it yet? Grab your phone (or old-school calendar or diary) and block out a couple of 30-minute sessions over a couple of weeks to get your money and your services working for the planet. Don't over-think it: the main thing is to switch to the provider that has the lowest emissions profile that your circumstances can manage and tell your old provider exactly why you're leaving. It's also worth checking in every year or so, as prices for climate-friendly services will very likely continue to fall.

* Greenpeace's Green Electricity Guide makes it easy to sort the polluters from the renewers in the electricity grid: https://greenelectricityguide.org.au
* Market Forces helps you judge your superannuation funds and banks so you can get your money out of fossil fuels:
  —www.marketforces.org.au/superfunds
  —www.marketforces.org.au/info/compare-bank-table
* Got shares beyond your super? Register your shareholdings with the Australasian Centre for Corporate Responsibility so you can help support shareholder resolutions when annual general meeting season rolls around: www.accr.org.au/shareholders
* 350.org Australia has long campaigned on Australian banks that lend to fossil fuel projects and companies; join here: https://350.org.au

SOLIDARITY NOW

Be a good ally to Aboriginal and Torres Strait Islander peoples by joining these groups, or donating today:

* Indigenous Peoples Organisation Australia: https://indigenouspeoplesorg.com.au
* Firesticks Alliance: www.firesticks.org.au
* First Nations Clean Energy Network: www.firstnationscleanenergy.org.au
* Original Power: www.originalpower.org.au
* Our Islands Our Home campaign: https://ourislandsourhome.com.au

* Regenerative Songlines Australia:
  www.regenerative-songlines.net.au
* Seed Indigenous Youth Climate Network:
  www.seedmob.org.au

WHERE DO AUSTRALIA'S EMISSIONS COME FROM?

It's handy to keep in mind the sources of the climate pollution we need to deal with: it can help you prioritise where and how you take action. Here's the Federal Government's inventory of the twelve months to the end of 2020, data that is compiled using emissions estimation rules adopted under the Paris Agreement.

THE MAIN SOURCES OF AUSTRALIA'S DOMESTIC CLIMATE POLLUTION

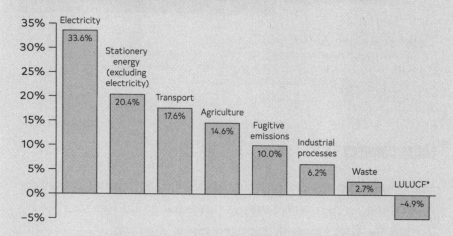

* LULUCF includes emissions associated with land use, land-use change and forestry.
Source: Australian Government, Department of Industry, Science, Energy and Resources, National Greenhouse Gas Inventory Quarterly Update: December 2020, www.industry.gov.au/data-and-publications/national-greenhouse-gas-inventory-quarterly-update-december-2020#annual-emissions-data-by-sector

AUSTRALIA'S FOSSIL FUEL EXPORTS EMISSIONS

Exports emissions are the emissions that result from fossil fuels that are dug up in Australia, sent overseas, and then used by other countries. Our exported emissions are far greater than our domestic emissions.

AUSTRALIA'S DOMESTIC VERSUS EXPORTED EMISSIONS

Australia's domestic emissions (Scope 1 and 2), 2019 calendar year*	529.3 megatonnes CO2e
Australia's exports emissions (Scope 3), 2019/20 financial year	1283 megatonnes CO2e

*As measured by the UNFCCC. Includes LULUCF.
Source: Climate Council; Scope 1 and 2 emissions sourced from Australian Government, Department of Industry, Science, Energy and Resources, Australian Energy Statistics, Table N, September 2021.

LOCAL CONNECT

Many seasoned advocates began their climate journey by talking to a friend or two and working out what to do together, or by joining the first group that looked like a good fit.

* Start a conversation on climate, with help from Climate For Change: www.climateforchange.org.au
* Join a local group of these legendary organisations:
  — Australian Conservation Foundation: www.climateforchange.org.au

— Friends of the Earth Australia: www.foe.org.au/members

— The Wilderness Society's Movement For Life: www.wilderness.org.au/work/movement-for-life

* There is likely to be a Landcare group near you; find it here: https://landcareaustralia.org.au/landcare-get-involved/findagroup

* Your local conservation council or environment centre can help direct you to local climate and environmental groups (and they also lead a bunch of critical climate campaigns you can join):

— Arid Lands Environment Centre: www.alec.org.au

— Cairns and Far North Environment Centre: https://cafnec.org.au

— Conservation Council SA: www.conservationsa.org.au

— Conservation Council of Western Australia: www.ccwa.org.au

— Environment Centre Northern Territory: www.ecnt.org.au

— Environment Tasmania: www.et.org.au

— Environment Victoria: https://environmentvictoria.org.au

— Hunter Community Environment Centre: www.hcec.org.au

— Nature Conservation Council of New South Wales: www.nature.org.au

—Queensland Conservation Council: www.queenslandconservation.org.au

* Find your local climate and sustainability group by postcode here (and there's a clickable map!): www.climatecrew.io

* The Next Economy supports communities across regional Australia to develop plans that reduce emissions while simultaneously creating good social and economic outcomes: https://nexteconomy.com.au

EVERYDAY ACTION

Lowering your impact is important, but wouldn't it be great to have a personal coach? Happily the One Small Step app is just that, so download it here: www.onesmallstepapp.com

---

THERE'S NO PLACE LIKE HOME

Making your home climate-friendly is a great thing to do if you're able: here are a few places to get you started:

* Renew provides independent, quality advice:
  https://renew.org.au
* My Efficient Electric Home group on Facebook is feast of tips and solutions:
  www.facebook.com/groups/996387660405677

Remember, if you're renting don't beat yourself up: campaign for healthier homes with Better Renting: www.betterrenting.org.au/healthy_homes_for_renters

WWF Australia has a handy carbon footprint calculator and loads of tips on how you can decarbonise everyday living: www.wwf.org.au/get-involved/change-the-way-you-live#gs.pmp4qb

When it comes to everyday campaigning, here's a few tips:

* Sharing something on social media you're furious about or pleased to see changing? Remember to tag your local politician by using the @ symbol and their profile name. Trust me, they notice.
* Speaking of engaging your politician (remember they represent you), a polite, heartfelt letter or phone call from you and your

friends will get your local MP listening. For tips on how to write letters, submissions or how to visit a pollie, head to The Commons library: https://commonslibrary.org/topic/lobbying-advocacy or The Change Agency: https://thechangeagency.org/resources—you'll find a heap of other resources on campaigning and organising there too. You can get transferred to any politician's office in Parliament House, Canberra, by dialling (02) 6277 7111.

* Check out Democracy in Colour, an organisation that works towards a society where the inherent worth, dignity and humanity of everyone is recognised: https://democracyincolour.org

* Furious about something no one's fixing yet? Start your own campaign with GetUp's handy tool https://me.getup.org.au or at Change.org www.change.org

WORK IT

Most of us are at work most of the time, so stack climate action into your 9 to 5.

* Work for Climate is the place to learn how to pitch climate action to your employer: www.workforclimate.org

* Professionals Advocating for Climate Action is a growing network with monthly meetings held online: www.professionalsforclimate.com.au

* Better Futures Australia is an alliance of companies, professional associations and individuals representing more than 7 million Australians, totalling more than $330 billion in GDP, assets and market capitalisation that your organisation can join: www.betterfutures.org.au

TEAM UP

These days there is a climate group for everyone. Get started here:

* AFL Players for Climate Action: www.aflp4ca.org.au
* Australian Parents for Climate Action: www.ap4ca.org
* Australian Marine Conservation Society:
  www.marineconservation.org.au
* Australian Religious Response to Climate Action:
  www.arrcc.org.au
* Australian Youth Climate Coalition: www.aycc.org.au
* Bushfire Survivors for Climate Action:
  www.bushfiresurvivors.org
* Climate and Health Alliance: www.caha.org.au
* CommsDeclare: https://commsdeclare.org
* Community Power Agency: https://cpagency.org.au
* Diplomats For Climate Action Now:
  www.diplomatsforclimate.org
* Doctors for the Environment Australia: https://dea.org.au
* Frontrunners: www.frontrunners.org.au
* Green Music Australia: www.greenmusic.org.au
* Groundswell Giving: www.groundswellgiving.org
* Healthy Futures: www.healthyfutures.net.au
* Jewish Climate Network: www.jcn.org.au
* Outdoors People for Climate: www.outdoorspeople.org
* RE-Alliance: www.re-alliance.org.au
* School Strike 4 Climate Australia:
  www.schoolstrike4climate.com
* Solar Citizens: www.solarcitizens.org.au

* Stop Adani: www.stopadani.com
* Surfers for Climate: https://surfersforclimate.org.au
* Tomorrow Movement: https://tomorrowmovement.com
* Women's Climate Congress: www.womensclimatecongress.com
* Women's Climate Justice Collective:
  www.facebook.com/WomensClimateJusticeCollective
* Veterinarians for Climate Action: www.vfca.org.au

Told you: there are heaps! For more climate organisations, visit the Climate Action Network Australia: www.cana.net.au/our_members

NERD OUT

Learn more on the climate challenge, and the systemic shifts we need in our energy, transport and land-use sectors here:

* The Australia Institute and Blueprint Institute are experts in climate and energy: https://australiainstitute.org.au and www.blueprintinstitute.org.au
* Beyond Zero Emissions breaks down the jobs and manufacturing potential Australia has on its doorstep: https://bze.org.au
* For the latest science on climate change, visit the Climate Council: www.climatecouncil.org.au
* Get trained up at Climate Reality: www.climatereality.org.au
* Find out more on the climate tech landscape; there's a tasty mix at Climate Salad: www.climatesalad.com
* Interested in decarbonisation pathways? ClimateWorks Centre is the place to go: www.climateworksaustralia.org

STILL FEELING FREAKED OUT?

If you're frozen with freak-out, you'll be taking climate action with one hand tied behind your back. Get some help.

* Psychologists for a Safe Climate has a list of climate-aware practitioners you can search: www.psychologyforasafeclimate. org/cap-directory
* The Good Grief Network has a bunch of resources and programs you can join: www.goodgriefnetwork.org
* The Work that Reconnects Network online course 'Active Hope' steps you through a process that participants find both healing and motivating: https://workthatreconnects.org/event/ free-online-course-active-hope-foundations-training/2022-02-19
* Britt Wray's excellent Gen Dread blog has a fantastic list of resources to work with climate emotions, built in collaboration with The All We Can Save Project: https://gendread.substack. com/p/resources-for-working-with-climate

ACKNOWLEDGEMENTS

It's impossible to acknowledge the decades of tireless work from First Nations peoples, campaigners, advocates, scientists, philanthropists, business leaders, politicians and community leaders who have made the choice to put a safe and healthy future for Australians first, so I will narrow my thanks to a few who helped this book arrive in the world.

First, to the 100-odd people who generously took time to share their stories for this project, the folks who made it into these pages and those who did not: as it turns out you can't pack all of Australia's climate action into a tiny 80,000 words. I've captured more of your projects and thoughts at **www.climateactionstartshere.com**

Deep appreciation goes out to the early advisers on this project: Anna Rose and Arielle Gamble from Groundswell Giving, Joan Staples, Amanda Tattersall, Anny Druett, Sally Gillespie, Nicole Thornton and the wonderful Liz Wade from Good Grief: your passion for this project was utterly inspiring and your feedback invaluable. Thanks to the smartest, funniest crew in the known universe—all at The Sunrise Project—and special thanks to my colleagues Miriam Lyons, Mark Wakeham, Sam La Rocca and John Hepburn for their unfailing support of this idea.

My thanks to all at Allen & Unwin, who took a risk on this first-time author, especially publisher Elizabeth Weiss for her wicked sense of humour, guidance and straight talking, Angela Handley for utter patience and perseverance, and Susan Keogh and Hilary Reynolds for perfecting the prose. Thanks to Andrew Griffiths for his advice in the early stages of pulling this idea together, and to Dr Rebecca Huntley for many hours of conversation on how Australians are feeling about climate change. Thanks to Jessica Bineth and Laura Brierley for research help at critical moments, and to Dr Cecily Moreton, who told me many times over that this book was more adventure than endeavour. You were right.

To Martin, Maeve and Illy, thank you for loads of work on this project, completed during lockdowns and across a summer holiday: the dinners made, chocolate and cups of coffee delivered, space and time for me to write carved out of our busy schedule, and your hugs and many reminders to breathe or to kick back in front of a movie.

Thanks also to my family. To my brothers and sisters and their partners, Trisha, Greg, Rachel, Matt, Emma, Amy, Dan, Shirley, Colin, Justine, Paul and Tsukimi, your interest and words of support made an enormous difference. To my parents, Janice and Peter, thanks for always supporting my latest out-there project. A special mention goes to my dad Peter, who worked his whole career in the coal power industry, keeping the lights on across Australia.

Thanks to my extended family, the Booderee camping crew, for growing older together and helping me remember to laugh and laugh. And to the women in my life who cheered on this project, especially Clare, Steph, Pip and Dominika, who kept me safely strapped into this roller-coaster ride. Dominika and Maya, many thanks for reading early drafts. Pip, thank you for reminding me to get out of my head and look up to appreciate the beauty of Illawarra's escarpment in all its seasons.

Most importantly, thanks and acknowledgement to the owners of Dharawal Country, land that was never ceded, land my family is privileged to call home. This place is a paradise that shines despite the scars it carries from heavy industry, urbanisation and racism. To elders past, present and future, we are indebted to you for protecting this place for thousands of generations past and to come. We stand in solidarity, always.